Fundamentals of Engineering

D1392106

Fundamentals of Engineering

R. L. Timings

LONGMAN

Addison Wesley Longman Limited
Edinburgh Gate, Harlow
Essex CM20 2JE, England
and Associated Companies throughout the world

© Addison Wesley Longman Limited 1998

The right of R. L. Timings to be identified as author of this work has been asserted by him in accordance with the Copyright, Designs and Patents Act 1988

All rights reserved; no part of this publication may be reproduced, stored in any retrieval system, or transmitted in any form or by any means, electronic, mechanical, photocopying, recording, or otherwise without either the prior written permission of the Publishers or a licence permitting restricted copying in the United Kingdom issued by the Copyright Licensing Agency Ltd, 90 Tottenham Court Road, London W1P 9HE.

First published 1998

British Library Cataloguing in Publication Data
A catalogue entry for this title is available from the British Library

ISBN 0-582-30583-7

Set by 32 in Times $9\frac{1}{2}/12$ and Frutiger
Produced by Addison Wesley Longman Singapore (Pte) Ltd
Printed in Singapore

Contents

Preface

Fundamentals of Engineering is the first book to be published in a new Longman series. Other titles in this series are:

- *Fundamentals of Mechanical Engineering*
- *Fundamentals of Fabrication and Welding*
- *Fundamentals of Electrical and Electronic Engineering*

These books have been written to provide the underpinning knowledge and understanding required by trainees working for the National Vocational Qualification (NVQ) in Engineering Manufacture (Foundation: Level 2) as part of their training programme; especially if they are involved in an EMTA *Modern Apprenticeship in Engineering Manufacture* scheme. These books closely follow the subject matter set out in City & Guilds' Document 2222 and guidance on the depth of treatment has been taken from the EMTA *Industry Standards of Competence* documents. EMTA are the lead body for the development of NVQs in all aspects of engineering.

Each chapter deals with a separate unit. For example, Chapter 1 of *Fundamentals of Engineering* deals with the content of Unit 003: *Establishing effective working relationships*. However, there is some overlap of subject matter between the units and, where this occurs, it is treated only once. Therefore, matters of behavioural attitudes towards safety are only introduced in Chapter 1 and are dealt with in detail in Chapter 8 (Unit 004: *General health and safety*).

The optional unit of *Heat treatment* has been included here since it is closely associated with the mandatory unit *Engineering materials*. Both units contain sections on materials testing and inspection and again, to avoid overlap, all materials testing and inspection processes have been included within Chapter 6: *Engineering materials*.

All the chapters end with a selection of exercises. These will help with assessing the trainee's performance criteria of the underpinning knowledge and understanding that is an essential part of his or her training. These end-of-chapter exercises are closely linked with, and help to reinforce, the workbook of NVQ Engineering Manufacture exercises written by David Salmon, and published in the Longman NVQ series.

R. L. Timings
1997

Acknowledgements

The author and publishers are grateful to the following for permission to reproduce copyright material:

The Royal Society for the Prevention of Accidents (RoSPA) for our Fig. 2.19; Myford Ltd for our Fig. 3.33; Taylor Hobson Ltd for our Fig. 4.7; Moore and Wright (James Neill Group) for our Fig. 4.17; TecQuipment Ltd for our Fig. 6.30(a); Samuel Denison Ltd (Denison Mays Group) for our Figs 6.30(b), 6.37, 6.41 & 6.42; Warwick Sign and Display Ltd for our Fig. 8.10; Silvaflame Ltd for our Figs 8.28 & 8.29.

Training Publications Limited for the reproduction and adaptation of some of their illustrations and text (our Figs 1.9, 8.1, 8.17, 8.18, 8.25, 8.34 & 8.35).

St. John Ambulance, St Andrew's Ambulance Association and the British Red Cross for text and illustrations in our Section 8.8 'First aid' (pages 293–8). Full information appears in the *First Aid Manual* published by Dorling Kindersley, price £9.99.

Extracts from British Standards Institution publications (our Figs 2.17, 2.18, 3.11, 3.12, 8.9, 8.10, 8.11, 8.12 & 8.13 and Tables 2.3, 4.3 & 4.4) are reproduced with the permission of BSI. Complete copies can be obtained by post from BSI Customer Services, 389 Chiswick High Road, London W4 4AL; Telephone: 0181 996 7000.

1 Establishing effective working relationships

When you have read this chapter, you should understand how to:

- Create and maintain effective working relationships with supervisory staff.
- Create and maintain working relationships with other people, members of the same working groups and other employees in the same organisation.

1.1 Basic relationships

Unless you have the misfortune to be shipwrecked on a desert island, you are unlikely to live and work in total isolation from other people. Let us suppose that you borrowed some money and set up a business in your garage making nesting boxes and bird tables. As a 'one-person' firm you may think you will not have to communicate with, or relate to, other people in the course of running your business. Not a bit of it! You will have to communicate with, and relate to, a surprisingly large number of people either through necessity or because it is the law. This is shown in Fig. 1.1.

In the first group (necessity) are such people as suppliers of raw materials and tools and equipment used in production. Also you have to deal with the customers who buy your products and the transport organisations who deliver your products to your customers. You also need a bank and since, from time to time, you may need an overdraft, it's as well to maintain good relationships with your bank manager. There is no law that says you have to have suppliers, customers and bankers but you would not get far in business without them.

There is no law that says you need a solicitor or an accountant. However, you require the former to draw up all the documents required when setting up the business and when problems arise with customers, suppliers and the local authority (e.g. noise complaints from the neighbours). You require the latter to audit your accounts, advise on financial matters, sort out your tax returns, make sure that you avoid over-payment of tax, and deal with the Customs and Excise officials over your VAT payments and returns. Therefore it is a *necessity* that you make every effort to maintain *good working relationships* with them.

In the second group (law) you have to communicate with such persons as Local Authority inspectors (planning officers, etc.), tax inspectors, VAT inspectors and Health and Safety inspectors. These people have the power of the legal system behind them so it pays to maintain *good working relationships* with them.

Fig. 1.1 *Structure of relationships*

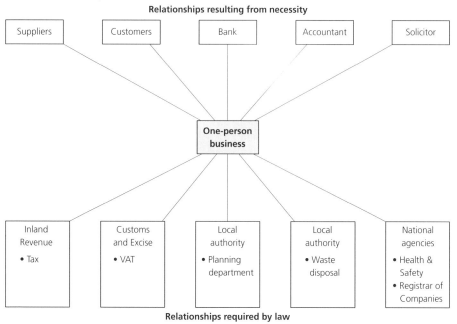

Relationships resulting from necessity

In our working lives we have constantly to relate to and communicate with other people. For example, we have to exchange technical data, implement management decisions and safety policy, and relate to other people within the company and also to people such as customers and buyers who work outside the company. In this section we are concerned mainly with the people with whom you will work on a daily basis; your workmates and your immediate supervisors and managers.

I hope I have made the point that no one can work in isolation even if he or she is the sole proprietor of a one-person firm. Certainly, you cannot work in isolation if you are an employer or an employee in a small, medium or large company. Like it or not, you are going to be one of a team. Like it or not, you are going to have to communicate, participate and cooperate. You are going to have to maintain *good working relationships*. When dealing with other people, you can adopt one of two possible attitudes. You can either *confront* them or you can *cooperate* with them, as shown in Fig. 1.2.

1.1.1 *Confrontation*

This is how the aggressive, bullying person works. The person who demands and threatens to get his or her own way. It may work in the short term as long as the aggressor has the whip hand. However, such aggressive bullies never win the respect of the people with whom they work. They can never rely upon the loyalty of the people they have continually confronted when a favour is required. It would be no good expecting 'good will' cooperation when an extra effort is required to complete an urgent order on time.

Fig. 1.2 *Confrontation and cooperation*

1.1.2 *Cooperation*

This is how sensible, civilised people work. They collaborate and help each other. In this way they gain respect for each other. This results in the development of efficient working relationships and efficient working practices. In an emergency every one can be relied upon to make a maximum effort and to help each other.

1.1.3 *'Reading' people*

As you become more experienced in dealing with people, you will realise that the most important skill is learning to 'read' their moods. You must be able to realise with whom you can have a joke with and with whom you can't. You need to know who only wants a 'yes or no' answer and who prefers to discuss a problem. You need to know when to be friendly and when to be aloof, when to offer a word of sympathy or advice and when to leave somebody alone to get over a bad mood, as shown in Fig. 1.3.

Fig. 1.3 *Reading people's moods*

1.2 Relationships with managers, supervisors and instructors

You are employed by the firm for whom you work, but you are responsible to your immediate superior. Depending on the structure of the company, your immediate superior may be an instructor, a charge-hand, a foreman or forewoman, a supervisor, a manager or, in a very small firm, the 'Boss'.

In a large company, there could be a whole management team above you who can offer a wide range of skills from which you can learn. Figure 1.4 shows a typical management structure for the training department of a large company. The structure will vary from firm to firm but, whatever the size of the company and the structure of its training facilities, it is always a good idea to find out what the structure is. You need to know who influences your training package, who actually trains you and who is responsible for your welfare, discipline and assessment. The change from the school environment to the adult working environment often poses unforeseen problems, it is essential to know who you should turn to when you need advice.

Fig. 1.4 *Training personnel structure*

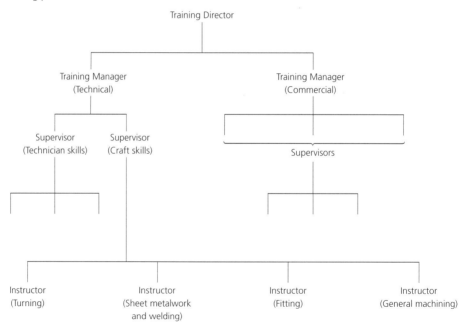

First and foremost, it is most important that you get on well with your instructor, your supervisor and your training manager. Each of them will require a different approach. This is not only because they are different people, but also because they have a different status and a different level of importance in the company. Let's now see how you can make a 'good impression' on these people and establish good working relationships with them. For example:

- Develop a habit of good time-keeping and regular attendance, even under difficult conditions.
- Be neat and tidy in your appearance.
- Keep your workstation neat and tidy and your tools and instruments in good condition.
- Keep your paperwork up to date, fill it in neatly and keep it clean in a plastic folder.
- File your paperwork systematically so that you can produce it for your instructor or your training manager on demand. 'Attention to detail' always makes a good impression.
- Be reliable so that people quickly find that they can depend upon you.
- Be conscientious: always try your hardest and do your best.
- Reasonable requests for information should be dealt with promptly, accurately and in a cooperative manner providing they do not unduly interfere with or interrupt your work.
- If responding to any request is going to take time and interrupt your work, or if it requires you to leave your working area, always seek permission from your supervisor or instructor before carrying out the request Always turn your machine off before leaving it or an accident may occur, as shown in Fig. 1.5.
- If you are in the middle of an intricate piece of work that requires your full concentration, don't just down tools, but ask politely if you may complete your task before actioning the request.
- No matter how tired you are or how inconvenient, trivial and unnecessary the request may seem to you, always try to be cheerful, helpful and efficient. **Never** answer in a surly, uncooperative, couldn't care less, any old time will do, manner.

Fig. 1.5 *Turn your machine off when you leave it*

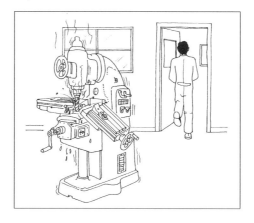

However, you are not a robot programmed to do as you are told without any reaction. Your relationships with other people, particularly your instructor, must be a dialogue of instruction and advice. If you are in doubt you must always discuss your problem with your instructor (see Fig. 1.6) until you are certain that you fully understand what you have to do.

Fig. 1.6 *When in doubt, **ask***

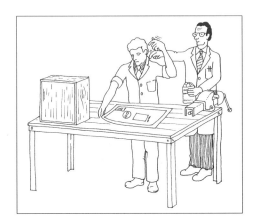

Your instructor is also there to help you with any personal problems you may have or problems with other people with whom you have to work. Your instructor wants to get to know you as a person so that he or she can get the best from you and help you to make a success of your training.

Remember, there is a time and place for everything, and should your instructor be talking to another trainee or his supervisor or manager, don't just barge in, as shown in Fig. 1.7. Either get on with another job and come back later or wait to one side respectfully until it is your turn. Be patient, on no account try to start work on a job or on a machine without instruction just because your instructor is busy and you are tired of waiting for him or her.

Fig. 1.7 *Don't push in, wait your turn*

1.3 Attitude and behaviour

1.3.1 *Attitude*

So far we have considered your general attitude to your instructor, supervisor and training manager. We now have to consider your attitude to your instructor and your work in more detail. You may have got away with 'giving cheek' to a teacher at school, but such an attitude would never be tolerated at work. You are a very new, very unimportant and very expendable member of the workforce. You know little or nothing about the skills of engineering so, if you are going to complete your training successfully and become a useful member of the company and of society as a whole, you've got a great deal to learn.

Your training is a major investment for your employer. Therefore employers need to train and employ reliable people who they can rely upon and who will give them a reasonable return for their investment in time and money. Those who demonstrate good attitudes are the most likely to succeed. It is no good being the most skilful apprentice or trainee if you are also the most temperamental. Whilst high levels of skill are important, so is consistency, reliability, loyalty and the ability to work in a team.

The greatest incentive to learning a trade is the earning power it gives you. To learn a trade you need the skilled help and advice of a lot of people. You must respect their skill and experience if you are to get their help and advice in return. However, unlike the character in Fig. 1.8, you do not have to grovel to show respect. There are many better ways of showing it and earning it.

Fig. 1.8 *You don't have to grovel to show respect*

Apart from the advice already given, here are some further suggestions:

- Dress in the way recommended by your company. Many firms provide smart overalls bearing the company logo. Do not turn up at work looking scruffy and wearing a long hair style. As you will find in the chapter on safety (Chapter 8), it not only gives a bad impression but it can also be very dangerous.

- For hygienic reasons change into clean overalls daily if possible. Dirty, oily overalls can cause serious hygiene and health problems. A tidy person has a tidy and receptive mind.
- Listen carefully to the instructions your instructor gives to you; particularly safety instructions. Never operate a machine or carry out a process if you are in doubt; always check again with your instructor.
- Keep a log of the operations you are taught and the work you do because your practical skill training has to be assessed in order for you to obtain your certification. Since you may have to present your log book at a future job interview, it is worth while spending some time on it. Keep it neat and clean in a plastic folder. Most firms expect their trainees to do this anyway, and provide standard log books.
- Show consistency, commitment and dedication in carrying out the tasks set for you. Work to as high a standard as you can and always try to improve your standards. Have pride in your work, you never know who is going to look at it. This applies equally to organisational tasks as it does to the production of components and assemblies.

1.3.2 *Behaviour*

In an industrial environment horseplay and fooling around infers reckless and boisterous behaviour such as pushing, shouting, throwing things and practical joking by a person or a group of persons. Engineering equipment is potentially very dangerous and 'playground' behaviour cannot be tolerated in an industrial environment. Behaviour will be dealt with in greater detail in Chapter 8.

In addition to the negative attitude to behaviour just described, there are positive attitudes. For example, keep your workstation clean and tidy, as shown in Fig. 1.9. Also, clean up any spillages immediately and ensure that the area where you are working is clear of swarf and other rubbish. Use the waste bins provided.

Fig. 1.9 *Good housekeeping*

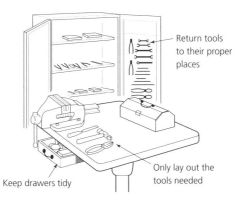

Return tools to their proper places

Keep drawers tidy

Only lay out the tools needed

1.4 Implementing company policy

Company policy may be set by the 'Boss' in a small company or it may be set by the board of directors in a large company. These people are not free agents and they have to abide by national and international laws and guidelines in setting out a strategy for the company. They have to consider the demands of the shareholders, and they are also responsible for the success, profitability and growth of the company upon which the job security and rewards of all who work for the company depend. For these reasons company policy should be understood and obeyed. In successful companies this is not an entirely autocratic process and there are various committee structures through which ideas from the shop floor can be fed back up the command chain to the senior management. This is particularly true for safety issues.

1.4.1 *Health and Safety*

This will be dealt with in detail in Chapter 8. All engineering and manufacturing companies are legally bound by the provisions of the Health and Safety at Work etc. Act of 1974, and other related legislation. The safety policy of your company must take on board all the requirements of such legislation.

Instructors and training managers have a vital part to play in fulfilling their obligations under such safety legislation and in anticipating and averting dangerous situations. Equally, by their manner in handling equipment and tools they must set a good example to their trainees and encourage attitudes of care and confidence.

It is equally important that trainees should understand how they fit into the system, what their duties are and to whom they should go for advice and help. Particularly, all trainees should know what to do in the event of an accident, how to carry out first aid, to whom and in what manner the accident should be reported and recorded and, in the event of a dangerous occurrence (even if it does not result in a notifiable accident), to whom and in what manner it should be reported.

1.4.2 *Communications*

No company can exist without lines of communications, both internal and external. Without internal lines of communication the company policy could not be communicated to the workforce, nor would the senior management know if their policies were being carried out. Figure 1.10 shows a typical management structure for a company.

This structure not only represent the lines of communication by which the senior management can ensure that decisions are passed down the line, it also represents a route by which messages and requests are sent back up to the various

Fig. 1.10 *Management structure*

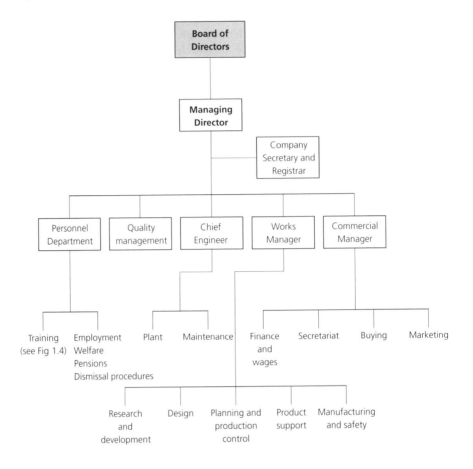

levels of management. These channels of communication are part of company policy; bypassing them can lead to confusion and friction between the parties concerned.

Wherever possible always use the standard forms provided when communicating within your company. This will result in your requests being treated seriously. Such forms may range from the stores requisition forms that you fill in daily, to job application forms and internal promotion application forms. Always follow laid-down procedures.

External lines of communication are equally important so that the company can communicate with its customers and suppliers. Market research, public relations and advertising are all essential to the success of a company and depend upon the use of suitable means of communication. This is why many companies employ firms of consultants specialising in these fields.

Verbal communications can take place via the telephone on a one-to-one basis or via meetings when information has to be given to a number of people at the same time. The advantage of verbal communication is that an instant response can be received and a discussion can take place. The disadvantage of verbal communication is its lack of integrity. Messages can be forgotten, they can be repeated inaccurately, or they can be misinterpreted.

There is an apocryphal story about the army in the days before field telephones and wireless and messages had to be sent by word of mouth. A commander sent a message down the line to headquarters: 'Send reinforcements, we are going to advance.' The message arrived at headquarters as: 'Send three and fourpence (old money) we are going to a dance!' All verbal messages should be backed up by written confirmation. This is why professional PR staff always back up an interview with a written and official 'press release'.

For accuracy, send a letter, fax, e-mail or a written memorandum in the first place. All firms of any standing use proforma documents such as letter paper with printed headings, official memorandum forms, official order forms, official invoice forms and despatch notes, and many other pre-printed documents. This ensures consistency of communication policy and saves time since only the details and an approved signature need to be added.

Nowadays electronics has speeded up internal and external communications and many firms are heading towards so-called 'paperless offices'. Increasingly, communications will be sent digitally by fax machines and e-mail. Business information is rapidly becoming available via the Internet. Files will be saved on disk instead of on paper in bulky filing cabinets.

No matter what system is used, communications cannot be fired off at random. There must be channels of communication that follow the chain of command, otherwise managers would not know what messages were being sent up to their superiors and back down to their subordinates – a recipe for chaos.

1.4.3 *Personal hygiene and dress*

Personal hygiene goes a long way to ensure good health and freedom from industrial diseases. It is also more pleasant for those who work with you if they do not have to put up with unpleasant and unnecessary body odours. As already mentioned, correct dress not only gives a good impression, it makes you look smart, feel smart and it helps you to avoid accidents. These matters will be dealt with in depth in Chapter 8, 'General Health and Safety (engineering)'.

Personal hygiene is most important. There is nothing to be embarrassed about in rubbing a barrier cream into your hands before work, about washing thoroughly with soap and hot water after work, about changing your overalls regularly so that they can be cleaned, or about wearing protective plastic or rubber gloves to protect your hands from chemicals and solvents. Personal hygiene can go a long way towards preventing skin diseases, both irritant and infectious. Your employer's safety policy should make recommendations on dress and hygiene and suitable protective measures should be provided.

1.4.4 *Dress*

For general workshop purposes, a *boiler suit* is the most practical and the safest form of protective clothing. However, such overalls must fit correctly and be kept in a good state of repair. If they are too long they cause you to trip and fall, particularly on stairways. Torn cuffs and sleeves can become caught in machinery.

Dirty and oil-soaked overalls are a major source of skin infection. This is why overalls should be regularly changed and cleaned. Finally, you must always wash your hands thoroughly before handling and eating any food, and when going to the toilet. If your hands are dirty and oily it is essential to wash them *before* as well as after.

Lightweight and unsuitable shoes should not be worn at work. Preferably, industrial safety shoes or boots should be worn. These will provide protection against:

- Severe puncture wounds resulting from treading on sharp objects.
- Crushed toes caused by heavy falling objects.
- Damage to your Achilles' tendon (this can be very serious) due to inadequate protection around your heel and ankle.

This is only a brief introduction to the subject of hygiene and projective clothing, and this important subject will be dealt with fully in Chapter 8.

1.4.5 *Recording and filing*

The need for keeping a training log and the need for using the standard forms supplied by your company has already been introduced. Nowadays most companies need their quality control system to be BS 5750 approved. This is because most of their customers will be so approved and will only be able to purchase their supplies from companies who are similarly approved. To trace the progress of all goods from supplier to customer, records must be kept and filed. The principles of quality control will be outlined in Chapter 2.

It is no use completing forms and keeping records unless they are properly filed. The success of any filing system depends upon the ease with which any documents can be retrieved on demand. If any file is removed from a filing system, a card must be inserted in its place stating who has borrowed the file and when. The file must be returned as soon as possible so that it does not become lost.

1.5 Creating and maintaining effective working relationships with other people

As has been stated previously, you cannot work in isolation. Sooner or later you have to relate to other people. In fact, most working situations rely upon teamwork.

1.5.1 *Positive attitudes*

Most of us have been to a high-street shop where the assistant finds serving us almost too much trouble, or where we are treated in a surly, couldn't care less attitude. How much better it is when we go to a shop to be greeted with a smile and where the assistant does his or her best to satisfy our requirements. It does not take much imagination to guess which shop we go back to next time.

Similarly, at work you should adopt a positive and constructive attitude to other people. This can be difficult when you are tired or the person you are relating to is off-hand, aggressive, demanding and asking for the near impossible. However, they are often under pressure themselves and allowances have to be made. Sometimes people are just out to annoy and provoke a confrontation. Try not to become involved. It is better to walk away from a quarrel than let it get out of hand. Always try to cool the aggressor down. Don't stand on your high horse and aggravate the situation or you can end up in a slanging match, as shown in Fig. 1.11. It does your reputation no good in the eyes of your workmates and, more importantly, in the eyes of your supervisor and the management.

Fig. 1.11 *Personal attitudes*

Sooner or later you are bound to come up against someone with whom you cannot get on. This may be a workmate or an instructor. Often, there is no apparent reason for this problem; it is simply a clash of incompatible personalities. If you cannot resolve the matter amicably yourself, don't leave the situation to deteriorate, but seek advice from the appropriate member of staff such as your supervisor or manager. He or she may be able to solve the problem which may involve you being moved to another section. Remember that, during your training, your personal attitudes and your ability to work as a team member are as much under scrutiny as the engineering products you produce.

1.5.2 *Teamwork*

Quite often you will have to work as a member of a team. This requires quite different skills in interpersonal relationships than when you are working on your own or under the guidance of your instructor. For example, consider the lifting of a large and heavy packing case when mechanical lifting gear is not available. Like any team, the lifting party has to have a team leader (captain). That person must have the respect and confidence of all the other members of the team because of his or her experience and expertise. The team should be picked from people who it is known can work together amicably and constructively. One member going his or her own way at a crucial moment could cause an accident and injury to other members of the team. Team lifting is discussed in detail in Chapter 8.

Although the team leader is solely responsible for the safe and satisfactory completion of the task, he or she should be sensible enough to consider comments and contributions from other members of the team. If you are a member of such a team and you think you have spotted a potential hazard in the job to be done, then it is your duty to draw it to the attention of the team leader. Eventually, however, discussion has to cease and the job has to be done. At this point the team leader has to make up his or her mind about how the job is to be tackled.

Remember: there can only be one team leader and his or her decisions must be obeyed by the rest of the team. The team leader should not take an active part in the exercise, but should stand back where he or she can see everything that is going on and, in the event of a potentially hazardous situation developing, be ready and alert to step in and correct the situation to prevent an accident.

1.5.3 *Personal property*

During a working lifetime most craftspersons acquire an extensive set of personal tools. Some may be bought and some may be made by the craftsperson personally. You will be mightily unpopular in any workshop if you borrow any of these tools without the owner's consent. The same applies to overalls or any other personal belongings. Although we have considered company policy, each and every workshop has a code of conduct all of its own. This is not written down, it is not company policy; it is a code of behaviour that has grown up over the years amongst the people working for that company, and woe betide anyone who disregards this code of conduct. However, if you respect it and obey it you will find that your relationships with your workmates and supervisors will be much more pleasant. You will receive more useful help and wise advice and will establish worthwhile friendships that can stand you in good stead throughout your working life.

EXERCISES

1.1   *Effective working relationships*
     (a)   You are engaged in an intricate machining operation when a colleague asks for your assistance. Explain how you should deal with this situation.

(b)   You are having difficulties in understanding and interpreting an engineering drawing and you want advice. Unfortunately your instructor is otherwise engaged in conversation with the departmental manager. Explain what would you do in this situation.

(c)   Your supervisor has directed you to help out with a team activity in another department. Explain how you would introduce yourself to the team leader and how you would try to relate to the other members of the team.

1.2   *Dress, presentation and behaviour*

(a)   Describe the dress code at your place of work or your training centre and explain why this dress code should be adhered to.

(b)   State **three** possible consequences of 'fooling about' in an engineering workshop.

(c)   Explain why you, as an engineering trainee, should:
(i)   adopt a short and neat hair style
(ii)   not wear dirty overalls, and change your overalls at frequent intervals
(iii)   write up your log book carefully and neatly, keep it in a plastic folder, and make sure it is available on demand for examination by your supervisor

(d)   Describe a situation where you need to wear special safety equipment.

1.3   *Instructions*

(a)   Draw an organisation chart to show the chain of command in your training centre or in a company with which you are familiar.

(b)   Upon receiving a verbal instruction, describe what you would do in order to ensure you have understood it correctly.

(c)   If a written instruction is unclear or badly printed, explain what you would do to avoid making a mistake in actioning the instruction.

1.4   *How to ask for help*

(a)   Describe a situation where your instructor might send you to some other person, such as a more senior colleague, for advice. Explain who that person might be in your training centre or company.

(b)   To avoid bothering your instructor when he or she is busy, describe:
(i)   the sort of practical assistance you might seek from a colleague
(ii)   the sort of information you might seek from a colleague
(iii)   the sort of advice you might seek from a colleague

(c)   State who you would approach for advice, and why you have chosen that person, in the following circumstances:
(i)   clarification of instructions or unclear advice from a colleague
(ii)   safe working practice concerning a new material that has been introduced into the workshop
(iii)   assistance in completing forms
(iv)   reporting personal injuries and accidents
(v)   discussing personal problems

(d) Give **one** example of the *correct approach* to another person when seeking that person's help or advice, and **one** *inappropriate approach* to another person when seeking that person's help or advice.

1.5 *How to give help when asked*
(a) List **five** important criteria that you must remember when giving help or advice to another person.
(b) Describe **three** situations when you should refuse to offer help or advice.
(c) Explain how you would try to make such a refusal without giving offence.

1.6 *Reporting deficiencies in tools, equipment and materials*
(a) Give **five** reasons why it is necessary to report deficiencies in tools, equipment and materials.
(b) Briefly describe the procedures used in your training centre or company for reporting defective tools, equipment and materials.

1.7 *Respect for other people's opinions and property*
(a) You may have to work with people whose values on work and life in general disagree with your own. Which of the following approaches should you adopt, and why?
     (i) argue aggressively with them
     (ii) respect their views, despite your personal reservations
(b) You are in a hurry and a long way from the stores. You know that your workmate has the equipment you need in his or her personal toolkit. What is the correct procedure for borrowing and returning such equipment?
(c) You are in a hurry to get home at the end of your shift. You are returning the tools you have been using to the stores. Should you check them and clean them or leave that to the store personnel to save time for yourself? Give reasons for your answer.

1.8 *Teamwork and cooperation*
(a) Why is it necessary to take the time and trouble to gain some knowledge and understanding of what other people do in your training centre or company, both within your department and in other departments? How could this lead to improved cooperation and team-work?
(b) How do some companies expand their trainees' and apprentices' insight into the work of other departments in the organisation?
(c) Give reasons for your answers to the following. When working in a team:
     (i) should you take part in the discussions concerning the work to be done?
     (ii) should you ask for clarification of matters you do not understand?
     (iii) from whom should you take your instructions?

1.9   *Difficulties in working relationships*

(a)   State **five** possible *causes* of difficulty that may arise in your relationships with your workmates and more senior staff.

(b)   With whom should you discuss such problems in the first place?

(c)   Describe the procedures that exist for formally reporting such difficulties in your training centre or company if you can get no satisfaction from (b) above?

2 Handling engineering information

When you have read this chapter, you should understand how to:

- Select information sources to undertake work tasks.
- Extract, interpret and evaluate engineering information.
- Record and process engineering information.

2.1 Selection of information sources

The need for clear communications that cannot be misinterpreted was introduced in Chapter 1. It is necessary, therefore, to select means of communication and information sources that ensure the correct information is provided and used. Engineering drawings are used to transmit and receive information concerning components to be manufactured and assembled. On the basis that it is easier to draw an elephant than to describe it, for many purposes drawings are much less likely to lead to misunderstanding and errors than verbal or written descriptions. Engineering drawings will be considered in detail in Chapter 3. However, some information has to be given in writing. For example:

- Manufacturing instructions such as the name of the parts to be made, the number required, any special finishes required and the date by which they are required.
- Technical data such as screw thread sizes, and manufacturers' recommended cutting speeds and feeds.
- Stock lists such as material sizes, standard 'bought-in' parts, and standard cutting tools.
- Training log-books.

Verbal instructions and telephone messages should be confirmed in writing or by fax. The latter is particularly useful if illustrations are involved. Think back to your school days. You often had to write essays on topics set by your teacher. If there were 25 pupils in the class, there would be 25 completely different treatments of the topic. Some essays would be short and lacking essential detail, whilst others would be long and rambling. Some essays would be easy to understand, whilst others would be obscure. The only thing that could be guaranteed is that they would all have been different. In industry and commerce all messages must be clear and concise, as there is not enough time to write down information in the form of essays. All information must be provided in a way that is:

- Easy to understand with no risk of errors.
- Complete, with no essential details missing.
- Quick and easy to complete.

These goals are best achieved by the use of standardised forms. By providing much of the information in the form of boxes that can be ticked, even the interpretation of handwriting that is difficult to read is overcome.

Manufacturing organisations are concerned with making the goods required by their customers at a price their customers are prepared to pay, and in delivering those goods in the correct quantities at the correct time. This involves teamwork within the organisations and close liaison with their customers and suppliers. This can only be achieved by the selection of efficient means of communications and the efficient handling of engineering information.

2.2 Interpretation of information (graphical)

There are many ways in which information can be presented and it is essential to select the most appropriate method. This will depend upon such factors as

- The information itself.
- The accuracy of interpretation required.
- The expertise of the audience to whom the information is to be presented.

Much of the information required for the manufacture of engineering products is numerical. This can be presented in the form of tables where precise information is required concerning an individual item. Sometimes, all that is required is a general overview of a situation that can be seen at a glance. In this case the numerical data is most clearly presented by means of graphs and diagrams. Graphs are a pictorial method of giving a clear and convenient representation of mathematical information, relationships and, frequently, trends. There are many different types of graph depending upon the relationship between the quantities involved and the numerical skills of the user for whom the graph is aimed. Let's look at some graphs in common use.

2.2.1 *Graphs*

Line graphs

Figure 2.1(a) shows a graph for the relationship:

$$N = \frac{1000S}{\pi \times d}$$

where N = spindle speed in rev/min
    S = cutting speed in m/min (15 m/min in this example)
    d = drill diameter in millimetres

Fig. 2.1 *Line graphs: (a) points connected by a smooth curve (points are related mathematically); (b) points connected by straight lines*

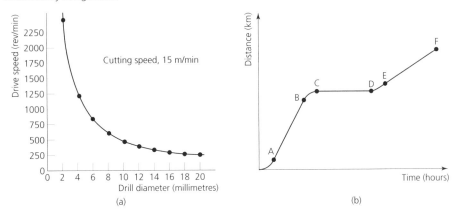

In this instance it is in order to use a continuous curve flowing through the points plotted. This is because the points plotted are related by a mathematical expression, and any value of N or d calculated from the expression will lie on the curve.

This is not true in every instance as shown by Fig. 2.1(b). This graph connects time and distance travelled for a vehicle:

- From A to B the distance travelled is proportional to the time taken; that is, the straight line indicates that the vehicle is travelling with a constant speed.
- The curved bit at the beginning of the line AB shows that the vehicle was accelerating from a standing start. The curved bit at the end shows that the vehicle slowed down smoothly to a stop.
- From C to D there is no increase in distance with time. The vehicle is stationary.
- From E to F the vehicle recommences its journey at a reduced speed since the line slopes less steeply.

In this graph the points are correctly connected by separate lines since each stage of the journey is unrelated to the previous stage or to the next stage. It would have been totally incorrect to draw a flowing curve through the points in this instance.

Histograms

Figure 2.2 shows the number of notifiable accidents which occur each year in a factory over a number of years. The points cannot be connected by a smooth, continuous curve as this would imply that the statistics follow some mathematical equation. Neither can they be connected by a series of straight lines. This would imply that, although the graph does not represent a mathematical equation, nevertheless the number of accidents increased or decreased continuously and at a steady rate from one year to the next. In reality the number of accidents are scattered throughout the year in a random manner and the total for one year is independent of the total for the previous year or the next year. The correct way to present this information is by a histogram as shown.

Fig. 2.2 *Histogram*

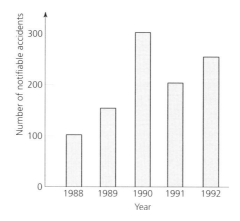

Bar charts

These are frequently used for indicating the work in progress and are used in production planning. An example is shown in Fig. 2.3.

Fig. 2.3 *Bar chart: × = scheduled completion date; ○ = actual completion date; ● = start delayed; shaded area = work completed to date*

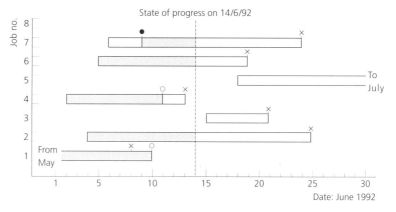

Ideographs (pictograms)

These are frequently used for presenting statistical information to the general public or other people who may not be expert in interpreting more formal graphical information. In Fig. 2.4 each symbol represents 1000 cars. Therefore in 1990, the number of cars using the visitors' car park at a company was 3000 (1000 cars for each of 3 symbols). Similarly, in 1991 the number of cars using the car park was 4000 and in 1992 the number had risen to 6000.

Fig. 2.4 *Ideograph (pictogram): number of cars using a car park each month*

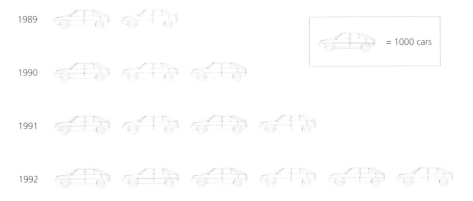

Pie charts

These are used to show how a total quantity is divided into its individual parts. Since a complete circle is 360°, and this represents the total, then a 60° sector would represent 60/360 = 1/6 of the total. This is shown in Fig. 2.5(a). The total number of castings produced by a machine company's foundry can be divided into the various categories of machine to be manufactured and can be represented by a pie chart, as shown in Fig. 2.5(b).

Fig. 2.5 *Pie chart*

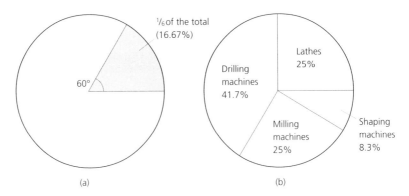

Flow charts

These are used to show a sequence of events and the order in which they occur. Figure 2.6(a) shows the sequence of operations for turning a hexagon head bolt on an automatic lathe, and Fig. 2.6(b) shows the sequence of operation for an assembly process.

Fig. 2.6 *Flow charts: (a) manufacturing flow chart; (b) assembly flow chart*

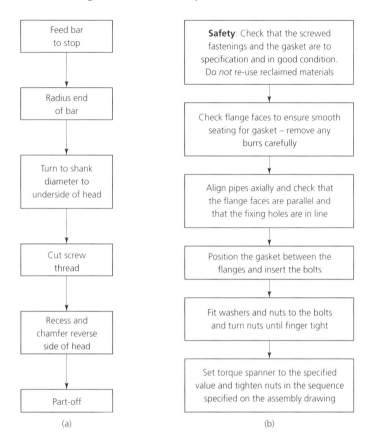

(a)

(b)

2.3 Interpretation of information (tables, charts and schedules)

2.3.1 *Manufacturers' catalogues*

Catalogues and technical manuals are an essential means of keeping up to date with suppliers' product lines. Also such catalogues and technical manuals usually include performance data and instructions for the correct and most efficient use of the products shown. For example, Figs 2.7 and 2.8 show two typical pages from a manufacturer's catalogue. Figure 2.7 shows the data required to choose suitable carbide inserts for turning tools for cutting cast iron. These are indexable, disposable inserts. They fit a standard tool shank and, as the cutting edge wears, they can be turned round (indexed) so that a new cutting edge becomes available. This enables a square insert to have four possible cutting edges. When all these edges become blunt, the insert is thrown away and a new one is installed. Since the inserts are mass produced it is easier and cheaper to discard a worn

Fig. 2.7 *Manufacturer's product data*

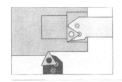

MEDIUM MACHINING OF CAST IRON

ISO/ANSI

K

T-MAX P

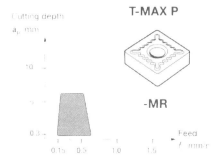

Cutting depth
a_p mm

10

1

0.3

0.15   0.5   1.0   1.5

Feed
f mm/r

-MR

Geometry -MR

Very strong cutting edge.

Secure and reliable geometry.

Excellent for sand inclusions and intermittence.

Application area: CNMG 120412-MR
a_p = 0.3 – 6.0 mm
f = 0.15 – 0.6 mm/r

For increased productivity

Coated silicon nitride.

Excellent wear resistance with reliable performance.

For best performance with ceramics use specially designed toolholders. See main catalogue.

-NGA / GC1690

FIRST CHOICE

-MR / GC4015

Tough edge and good wear resistance

High consistent performance.

For increased security

Interrupted cuts.

Very strong cutting edge and good wear resistance.

-MR / GC4025

Insert grades

Wear resistance

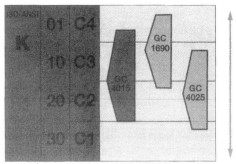

ISO/ANSI

K

01 C4

10 C3

20 C2

30 C1

GC 1690

GC 4015

GC 4025

Toughness

Grey cast iron, high tensile, HB 260

Cutting data/grade		Feed f_n mm/r		
CMC 08.2		0,2	0,35	0,5
GC1690	v_c m/min	635	515	444
GC4015	v_c m/min	215	170	140
GC4025	v_c m/min	205	160	135

Fig. 2.8 *Manufacturer's technical data*

ORDERING CODE	Productivity GC1690	Versatility GC4015	Security GC4025	COROKEY CUTTING DATA, CMC 08.2		
Toolholders, pages 34 – 36. Double sided				Starting value (range)		
				a_p mm	f_n mm/r	v_c m/min
CNGA 12 04 08 T02520	☆			3.6 (0.1 – 6.0)	0.20 (0.15 – 0.36)	635 (670 – 520)
CNMG 12 04 08-MR		★	☆	3.5 (0.3 – 6.0)	0.35 (0.15 – 0.50)	170 (235 – 140)
CNGA 12 04 12 T02520	☆			3.6 (0.1 – 6.0)	0.30 (0.15 – 0.55)	560 (670 – 425)
CNMG 12 04 12-MR		★	☆	4.0 (0.3 – 6.0)	0.40 (0.15 – 0.60)	160 (235 – 125)
CNGA 12 04 16 T02520	☆			3.6 (0.1 – 6.0)	0.40 (0.15 – 0.60)	495 (670 – 400)
CNMG 12 04 16-MR		★	☆	4.0 (0.3 – 6.0)	0.45 (0.20 – 0.70)	150 (215 – 115)
CNMG 16 06 08-MR		★	☆	3.5 (0.25 – 8.0)	0.35 (0.15 – 0.50)	170 (235 – 140)
16 06 12-MR		★	☆	4.0 (0.3 – 8.0)	0.40 (0.15 – 0.60)	160 (235 – 125)
CNMG 19 06 12-MR		★	☆	4.0 (0.3 – 10.0)	0.40 (0.15 – 0.60)	160 (235 – 125)
CNGA 19 06 16 T02520	☆			5.7 (0.1 – 9.5)	0.40 (0.15 – 0.60)	495 (670 – 400)
CNMG 19 06 16-MR		★	☆	5.0 (0.3 – 10.0)	0.45 (0.20 – 0.70)	150 (215 – 115)
DNGA 15 04 08 T02520	☆			4.5 (0.1 – 7.5)	0.20 (0.15 – 0.36)	635 (670 – 520)
DNMG 15 04 08-MR		★	☆	3.5 (0.3 – 6.5)	0.35 (0.15 – 0.50)	170 (235 – 140)
DNGA 15 04 12 T02520	☆			4.5 (0.1 – 7.5)	0.30 (0.15 – 0.55)	560 (670 – 425)
DNMG 15 04 12-MR		★	☆	4.0 (0.3 – 6.5)	0.40 (0.15 – 0.60)	160 (235 – 125)
DNMG 15 06 08-MR		★	☆	3.5 (0.3 – 6.5)	0.35 (0.15 – 0.50)	170 (235 – 140)
15 06 12-MR		★	☆	4.0 (0.3 – 6.5)	0.40 (0.15 – 0.60)	160 (235 – 125)
15 06 16-MR		★	☆	4.0 (0.3 – 6.5)	0.45 (0.20 – 0.70)	150 (215 – 115)
SNGA 12 04 08 T02520	☆			3.6 (0.1 – 6.0)	0.20 (0.15 – 0.36)	635 (670 – 520)
SNMG 12 04 08-MR		★	☆	3.5 (0.3 – 6.0)	0.35 (0.15 – 0.50)	170 (235 – 140)
SNGA 12 04 12 T02520	☆			3.6 (0.1 – 6.0)	0.30 (0.15 – 0.55)	560 (670 – 425)
SNMG 12 04 12-MR		★	☆	4.0 (0.3 – 6.0)	0.40 (0.15 – 0.60)	160 (235 – 125)
SNGA 12 04 16 T02520	☆			3.6 (0.1 – 6.0)	0.40 (0.15 – 0.60)	495 (670 – 400)
SNMG 12 04 16-MR		★	☆	4.0 (0.3 – 6.0)	0.45 (0.20 – 0.70)	150 (215 – 115)
SNMG 19 06 12-MR		★	☆	4.0 (0.3 – 10.0)	0.40 (0.15 – 0.60)	160 (235 – 125)
19 06 16-MR		★	☆	5.0 (0.3 – 10.0)	0.45 (0.20 – 0.70)	150 (215 – 115)
TNGA 16 04 08 T02520	☆			4.8 (0.1 – 8.0)	0.20 (0.15 – 0.36)	635 (670 – 520)
TNMG 16 04 08-MR		★	☆	3.5 (0.3 – 6.0)	0.35 (0.15 – 0.50)	170 (235 – 140)
TNGA 16 04 12 T02520	☆			4.8 (0.1 – 8.0)	0.30 (0.15 – 0.55)	560 (670 – 425)
TNMG 16 04 12-MR		★	☆	4.0 (0.3 – 6.0)	0.40 (0.15 – 0.60)	160 (235 – 125)
TNMG 22 04 08-MR		★	☆	3.5 (0.3 – 8.0)	0.35 (0.15 – 0.50)	170 (235 – 140)
22 04 12-MR		★	☆	4.0 (0.3 – 8.0)	0.40 (0.15 – 0.60)	160 (235 – 125)
22 04 16-MR		★	☆	4.0 (0.3 – 8.0)	0.45 (0.20 – 0.70)	150 (215 – 115)
WNMG 08 04 08-MR		★	☆	3.0 (0.3 – 4.0)	0.35 (0.15 – 0.50)	170 (235 – 140)
08 04 12-MR		★	☆	3.0 (0.3 – 4.0)	0.40 (0.15 – 0.60)	160 (235 – 125)

★ = First choice with CoroKey cutting data.
☆ = Alternative. Cutting speed recommendations, see table on previous page.

Ordering example: 10 pieces CNMG 120412-MR 4015

insert and use a new one than it is to try to regrind an old one. Further, they are made to high standards of dimensional accuracy and quality. This enables a new insert to be installed in the tool shank and for cutting to recommence without having to reset the machine. The performance of the replacement insert will be identical to the one discarded.

The original catalogue pages are easier to interpret since they are colour coded and larger than the reproductions in this book. The data shown in Fig. 2.7 provides information about the insert itself. This information is only shown as an example. It is not complete and needs to be read in conjunction with other pages in the catalogue. Figure 2.8 provides cutting data for optimum performance of the chosen insert, in terms of:

- The shape of the insert, the range of types available, and their catalogue number for ordering.
- The depth of cut (a_p mm) and the feed/rev (f_n mm/rev).
- The column headed v_c m/min is the cutting speed in metres per minute. This can be converted to spindle speed in rev/min by the expression:

$$N = \frac{1000\,s}{\pi \times d}$$

where N = spindle speed in rev/min
  S = cutting speed in m/min
  d = diameter of the workpiece before cutting

2.3.2 *British and European Standards*

At the start of the Industrial Revolution there was no standardisation of components. Every nut and bolt was made as a fitted pair which were not interchangeable with any other nut and bolt. Imagine finding a box full of nuts and bolts of seemingly the same size and having to try every nut on every bolt until you found which nuts fitted which bolts (see Fig. 2.9). No wonder that screwed fasteners were the first manufactured goods to be standardised, although, initially, only on a national basis.

Fig. 2.9 *Standardisation saves time*

Modern industry requires a vast range of standardised materials and components to provide the interchangeability required for international trading and uniformity of quality. Initially this work was carried out by such organisations as the British Standards Institution (BSI) in the UK, by DIN in Germany, and by ANSI in America. Since 1947, the International Standards Organisation (ISO) has been steadily harmonising *national standards* and changing them into *international standards* in order to promote international trading in manufactured goods. The aims of standardisation, as defined by the BSI, are:

- The provision of efficient communication amongst all interested parties.
- The promotion of economy in human effort, materials and energy in the production and exchange of goods through the mass production of standardised components and assemblies.
- The protection of consumer interests through adequate and consistent high quality of goods and consumer services.
- The promotion of international trade by the removal of barriers caused by differences in national practices.

In order to achieve these aims, the BSI, in collaboration with the ISO, publishes standards, codes of practice and glossaries as follows:

- *Standard specifications* These specify the requirements for particular materials, products or processes, and they also specify the method for checking that the requirements of the standard have been met.
- *Codes of Practice, Guides, etc.* These are recommendations for good practice. They aim to assist in ensuring the provision of goods and services which are safe and of consistent high quality.
- *Glossaries* To ensure that the Standards and Codes of Practice are correctly interpreted, glossaries are available. These are lists of specialist and technical words and phrases together with their meaning in the context of the Standards and Codes of Practice.

There are five categories of standardisation which apply to the engineering industry and these can be summarised as follows:

- Terminology and symbols (e.g. drawing symbols and conventions).
- Classification and designation.
- Specifications for materials, standard products and processes.
- Methods for measuring and testing, sampling and analysing.
- Recommendations on product or process applications and codes of practice.

2.3.3 *Production schedules*

These are usually in the form of bar charts or computer listings. The former will show the planned start and finish dates for various jobs and the machines onto which they are to be loaded. The actual progress of the jobs is superimposed on the ideal schedule so that any 'slippage' in production, and the reason, can be seen at a glance so that remedial action can be taken and, if necessary, the customer advised of possible delay. An example was shown in Fig. 2.3.

Computer listings of production schedules and stock balances are updated regularly (on a daily basis) so that the sales staff of a company know what components and assemblies are in stock, and how soon new stock should be available if a particular item has sold out. A typical example is shown in Fig. 2.10.

Fig. 2.10 *Typical stock list*

Batch quantity:	**100**			Description:			
Order point:	**56**			**Bearing block**			
Buffer stock:	**40**			Part no.:			
Maximum stock:	**160**			**3890 A**			
Minimum stock:	**30**						
Date	In	Out	Balance	Date	In	Out	Balance
3.6.91	98	–	98	2.9.91	–	12	36
10.6.91	–	12	86	9.9.91	–	2	34
17.6.91	–	16	70	16.9.91	104	–	138
24.6.91	–	14	56	23.9.91	–	12	126
1.7.91	–	14	42	30.9.91			
8.7.91	–	4	38				
15.7.91	–	14	24				
22.7.91	102	–	126				
29.7.91	–	16	110				
5.8.91	–	16	94				
12.8.91	–	14	80				
19.8.91	–	16	64				
26.8.91	–	16	48				

Significant dates
24.6.91 – Re-order
15.7.91 – Minimum breached; investigate
26.8.91 – Re-order

2.3.4 *Product specifications*

In addition to scheduling the work that is to be done and when, it is also necessary to issue full instructions to the works concerning the product to be made – that is, a *product specification* must be issued. For example, let us consider a car production line. It is set up to produce a continuous flow of a particular type of car. However, within that basic work pattern there are many variations. For example, some will have one colour and type of paint finish, others will be different; some will have one trim, others will have another; some will have power steering, others will not, and so on. Therefore, each car built will have a product specification, so that the customer will get the car he or she has chosen.

On a simpler basis is the *works order* issued in a batch production or in a jobbing workshop. This provides the information needed to manufacture a batch of components. An example of such a works order form is shown in Fig. 2.11.

Fig. 2.11 *Typical works order form*

ABC Engineering Co. Ltd		Job No.
Date issued	Date required	
Component		
Drawing numbers		
Quantity		
Material size	Type	Quantity
Tooling		
Finish/Colour		
Date commenced	Date finished	Operator
Inspection report		Inspector
Special requirements		
Destination		Authorised by

The example shown provides the following information:

- It identifies the component to be made.
- It identifies the drawings to be used.
- It states the quantity of the product to be made.
- It specifies the material that is to be used.
- It specifies any special jigs, fixtures, tools and cutters that will be needed and their stores location.
- It specifies any heat-treatment and finishing process that may be required.
- It specifies the issue date for the order and the date by which it is required.
- It specifies the destination of the job (stores, inspection department, etc.).
- It includes any special variations required by a particular customer.
- It identifies the personnel employed in the manufacture and inspection of the job.

- It carries the signature that gives the managerial authority for the work to be done.
- It provides room for the actual dates to be inserted when the job was commenced, and when it was finished.

You will notice that all this information is entered on a standard form. This saves time in issuing the information. It is much easier to fill in the blanks than have to write out all the information from scratch. It is also easy to see if a 'box' is blank. This would indicate that a vital piece of information is missing. It is also easier for the person doing the job to see exactly what is required since the same sort of information always appears in the same place on the form each time.

2.3.5 *Reference tables and charts*

There are a number of 'pocket books' published for the different branches of engineering. A typical 'pocket book' for use in manufacturing workshops would contain tables of information, such as:

- Conversion tables for fractional to decimal dimensions in inch units, and conversion tables for inch to metric dimensions.
- Conversion tables for fractional (inch), letter, number and metric twist drill sizes.
- Standard screw thread and threaded fastener data tables.
- Tables for spacing holes around pitch circles as an aid to marking out.
- Speeds and feeds for typical cutting tool and workpiece material combinations for different processes.

This list is by no means exhaustive but just a brief indication of the sort of useful data provided. A typical table of data is shown in Fig. 2.12.

In addition, many manufacturers produce wall charts of similar data as it affects their particular products. These are not only more convenient for the user than having to open and thumb through a book with oily hands, but they are also good publicity for the manufacturers who issue them.

2.3.6 *Drawings and diagrams*

It has already been said that it is easier to draw an elephant than describe it. Engineers use drawings and diagrams widely to communicate with each other and with the public at large. The type of drawing or diagram will depend upon the audience it is aimed at and their ability to interpret such information correctly. The creation and interpretation of engineering drawings is considered in detail in Chapter 3 of this book. The drawings that you are most likely to meet in engineering workshops are:

- *General arrangement drawings*   These show all the components assembled together. They also list the drawing numbers for the *detail drawings* required for the manufacture of the various components that go to make up the assembly. In addition, they also list the ready-made 'bought-in' components such as nuts, bolts, washers and dowels together with catalogue data to aid the ordering of such a component. An example is shown in Chapter 3, Fig. 3.31.

Fig. 2.12 *Typical data sheet*

Metric series	CUTTING SPEEDS Approximate							Metric series
ft/min	30	40	50	60	70	80	90	100
m/min	9	12	15	18	21	24	27	30
diam/mm	Revolutions per minute							
0.5	5817	7756	9695	11634	13573	15512	17451	19390
1.0	2909	3878	4847	5817	6786	7756	8725	9695
1.5	1942	2589	3237	3884	4532	5179	5826	6474
2.0	1456	1942	2427	2912	3397	3883	4369	4854
3.0	970	1294	1617	1940	2264	2587	2911	3234
4.0	728	970	1213	1455	1698	1940	2183	2425
5.0	582	777	970	1164	1359	1553	1747	1941
6.0	485	647	808	970	1132	1294	1455	1617
7.0	416	555	693	832	970	1109	1248	1386
8.0	364	485	606	728	849	970	1091	1213
9.0	324	431	539	647	755	962	970	1078
10.0	291	388	485	582	679	776	873	970
11.0	265	353	441	529	617	706	794	882
12.0	243	324	404	485	566	647	728	808
13.0	234	299	373	448	522	597	672	746
14.0	208	277	346	416	485	554	623	693
15.0	194	259	323	388	453	517	582	647
16.0	182	243	303	364	424	485	546	606
17.0	171	228	285	342	399	456	513	571
18.0	162	216	269	323	377	431	485	539
19.0	153	204	255	306	357	408	459	511
20.0	146	194	242	291	340	388	436	485
21.0	139	185	231	277	323	370	416	462
22.0	133	177	220	265	309	353	397	441
23.0	127	169	211	253	295	337	380	422
24.0	121	162	202	242	283	323	364	404
25.0	117	155	194	233	272	310	349	388

- *Detail drawings* As their name implies, these drawings show all the details of shape and dimension for making the individual components. They also specify the material, any heat-treatment required and any special finishes such as electroplating. An example is shown in Chapter 3, Fig. 3.32.
- *Exploded views* These are usually included in service manuals for machine tools and other mechanisms. They show how the parts are arranged relative to each other to aid removal and reassembly. They are usually listed and numbered to ease the ordering of spares that may be required during maintenance and repair. An example is shown in Chapter 3, Fig. 3.33.

2.3.7 *Other members of the workforce*

When you start out in engineering you will be working alongside older and more experienced persons. Listen to their advice. They have trodden the path before you; they have made mistakes and learned the hard way; they know many 'tricks of the trade' that can make a job easier and the result of your efforts better. Don't be too proud to ask for their help and advice. Most are usually only to pleased to help and are often flattered to be asked. However, pick your moment. Don't interrupt someone who is concentrating on a particularly difficult job – your knowledge of shop-floor terminology is liable to be suddenly and extensively enlarged! Further, treat such advice with caution. Many older and experienced workers have also picked up a lot of bad habits over the years. Check their advice with your instructor or supervisor before using it.

2.4 Evaluating engineering information

Keep alert for errors in the information given. Suppose you have made several batches of a component from stainless steel and suddenly the works order form specifies silver steel. Is this a genuine change or a clerical error? The manager may have signed it, but he is a very busy person and may have missed the error. Therefore, check with your supervisor before starting the job. Better to be sure than sorry.

If standards are referred to, check that the issue on the shop floor is up to date. Standard specifications and EU regulations change frequently. Out-of-date editions should be withdrawn immediately and the latest edition issued. However, it is surprising how long an out-of-date copy can continue to circulate before someone spots it and destroys it.

Don't take the information in tables of data and information charts for granted. No matter how carefully edited, errors still creep in. If some piece of information is out of the ordinary, cross-check with another set of tables. Alternatively seek the opinion of your instructor or supervisor. Engineering equipment is expensive and potentially dangerous. A misplaced decimal point in a table of cutting speeds and feeds could result in a tool or machine being overloaded, with disastrous results. An abrasive wheel run over speed could cause it to burst with potentially lethal results.

2.5 Recording and processing engineering information

The need for, and importance of, accurate record keeping is increasing all the time in nearly all the areas of company activity. Let us now look at some of the more important aspects that affect all employees.

2.5.1 *Quality control*

Quality control now affects nearly all manufacturing companies both large and small. This is because a firm that wants to sell its goods to a BS 5750 approved firm, must itself be BS 5750 approved and, in turn, obtain its supplies from BS 5750 approved sources. So, what is BS 5750 (ISO 9000)? In the UK the British Standard for Quality Assurance is BS 5750 (ISO 9000: 1987). The definition of quality upon which this standard is based is in the sense of '*fitness for purpose*' and '*safe in use*', and that the product or service has been designed to meet the needs of the customer.

A detailed study of quality control and total quality management is beyond the scope of this book. However, if you are employed in the engineering industry it is almost inevitable that you are employed in a company that is BS 5750 approved, and this will influence your working practices. A key factor in this respect is '*traceability*'. Therefore BS 5750 is largely concerned with documentation procedures. All the products needed to fulfil a customer's requirements must be clearly identifiable throughout the organisation. This is necessary in order that any part delivered to a customer can have its history traced from the source from which the raw material was purchased, through the stages of manufacture and testing, until it is eventually delivered to the customer. This is shown diagrammatically in Fig. 2.13. The need for this traceability might arise in the case of a dispute with a customer due to non-conformity with the product specification, for safety reasons if an accident occurs due to failure of the product, and for statutory and legal reasons. For these reasons, like everyone else

Fig. 2.13   *Quality control chain. (Each stage is a customer of the previous stage and a supplier to the next stage, i.e. Department 'A' is a customer of the Stores and a supplier to Department 'B')*

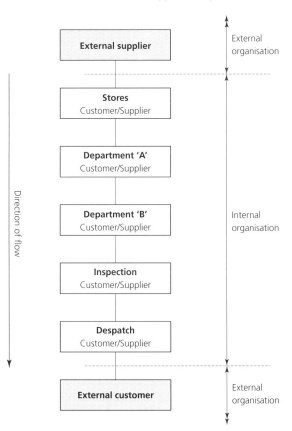

involved, your contribution to the above chain of events has to be accurately recorded and the records kept indefinitely. Otherwise the company's goods may not be certified and acceptable to the customer.

2.5.2 *Health and safety*

This is discussed in detail in Chapter 8 of this book. Here, also, record keeping is essential. Some of the documents and data that need to be maintained are:

- The accident register.
- The regular inspection and certification of lifting tackle and pressure vessels (boilers and compressed air receivers).
- Dates of fire drills and the time taken to evacuate the premises: records of visits to the premises by local authority fire and safety officers and their reports.

2.5.3 *Legal and financial reasons*

Registered companies need to keep legal records to comply with the Companies Act. They must publish their Memorandum and Articles of Association and lodge a copy at Companies House when the company is set up. They must keep accurate minutes of all meetings of the Board of Directors and make annual returns, including a current list of directors and other information, immediately following each Annual General Meeting (AGM) of the company.

Similarly, it is important that a company keeps accurate and complete financial records so that it can keep its costs under control and ensure that a profit is made. It also needs these records to satisfy the accountants when they make their annual visit to audit the accounts and draw up the balance sheet as required by the directors and shareholders, as well as the tax authorities.

2.6 Methods of record keeping

2.6.1 *Computer files*

Most records are now kept on computer files in the form of magnetic disks, magnetic tapes and optical disks (compact disks). These are easily destroyed by fire, theft and computer viruses (bugs). For this reason such data should be regularly backed up so that, in an emergency, the data can be reinstated with the minimum of down time and loss of business. Most modern computers automatically 'save to disk' at preset intervals (say every 20 minutes) that the computer is in use. Also, at the end of the day all computer transactions should be saved by 'tape streaming' so that a total back-up copy of all the previous day's transactions is available. Such back-up copies should be kept in a burglar-proof and fire-proof safe.

2.6.2 *Microfilm and microfiche*

Paper records are bulky and easily lost and destroyed. Forms and technical drawings can be easily copied onto microfilm or microfiche systems. These can store large quantities of information photographically in a small space. Such material can be conveniently catalogued and, when required, can be read through a suitable viewer or enlarged to its original size in the form of a photographic print.

2.6.3 *Log books*

These are used for various purposes such as recording the maintenance history of machine tools. The testing and inspection of equipment such as lifting gear, pressure vessels and fire extinguishers. Log books are also kept by lorry drivers and sales representatives to maintain records of their journeys.

Training log books

For young trainees, one of the most important documents to be kept is your *training log book*.

The format of log books varies from one training establishment to another. No matter what format is chosen, your log book should:

- Record the training you have undergone.
- Show details of the exercises you have undertaken and how you carried them out.
- Show how successfully you have completed each exercise.
- Show your instructor's comments on your performance and his or her signature verifying the entry.

Figure 2.14 shows a typical page from a trainee's log book. Before you can commence the work you will need to draw up an operation planning sheet. An example of a suitable planning sheet to lie alongside the job entry is shown in Fig. 2.15. This will enable you to plan the operations in a logical sequence in advance of starting the job so that you keep the number of settings and operations to a minimum and don't leave yourself with nothing for the machine to hold onto when you come to the final operation. It also enables you to requisition all the tools, measuring instruments and materials you will require before you commence and avoid having to make repeated and time-wasting journeys to the stores.

Your training log book is a most important document, therefore you must take great care in writing it up neatly and accurately. It provides a certified record of your training and the exercises you have completed. It will provide the evidence (evidence indicator) that will be required by the body responsible for the issue of a satisfactory training certificate on completion of your training. At some time in the future, a well-kept and presented training log book can be a key factor in a successful job interview. Another document that should be kept with your training log book is your daily training diary. A typical entry is shown in Fig. 2.16.

Fig. 2.14 *Typical page from trainee's log book*

JOB ENTRY

TOOLS USED	SAFETY POINTS
3 - Jaw chuck , Drill chuck	Do not leave machine running
Centre drill (B53) , Twist drills	unattended .
ø20 mm and ø9·5 mm . set of taps	Clean up any spilt coolant .
M12 x 1·75 . Facing tool , bar turning	Never leave chuck key in chuck.
tool , boring tool . Revolving centre .	Keep work area tidy .
Micrometers 0 – 25 mm & 25 – 50 mm.	Stop machine to take measurements.
Vernier caliper , Rule & odd-leg calipers .	Do not remove swarf with
D. T. I. and stand.	fingers .

(ENTER AGAINST MODULE SPECIFICATION OR TRAINING SYLLABUS)

Turning exercise . Drg 1/87

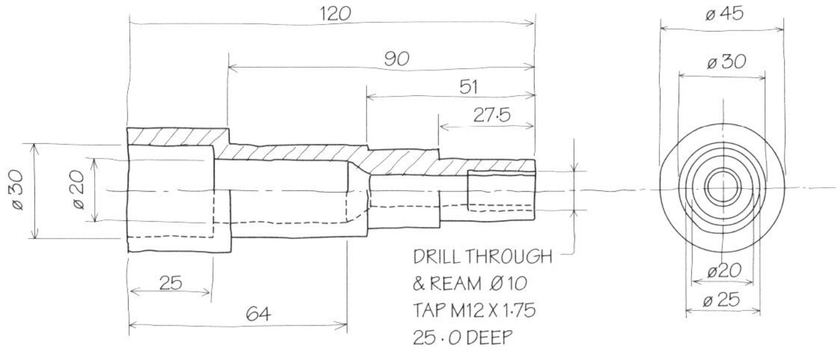

DESCRIPTION OF JOB WITH WORKS OR DRAWING No.

JOB No. 10

120
90
51
27·5
ø 30
ø 20
25
64

ø 45
ø 30
ø20
ø 25

DRILL THROUGH
& REAM Ø 10
TAP M12 X 1·75
25 · 0 DEEP
DIMENSIONS IN MILLIMETRES
GEN . TOL ± 0·25 mm
MATL. BDMS ø 50

PROGRESS REPORT

QUALITY OF WORK	**B**	INTEREST IN WORK	**B**	INITIATIVE	**C**	TIMEKEEPING	**A**
SPEED OF WORK	**D**	SAFETY OF WORK	**B**	CONDUCT	**B**	LEARNING RATE	**C**

SUPERVISOR'S REMARKS: *Produces good quality work but needs to work faster without loss of accuracy.*

TRAINING OFFICER'S REMARKS: *Anne continues to make satisfactory progress but needs to meet time targets*

LOG BOOK MARK **7/10**   SIGNED:

MARKING STANDARD   A = EXCELLENT   B = GOOD   C = AVERAGE   D = POOR   E = UNSATISFACTORY

Fig. 2.15 *Operation sheet (log book)*

NAME Joan Smith	DATE 7/1/97	DRG No. 2/12	EXERCISE 16		
OP No.	DESCRIPTION OF OPERATION		SPEED (rev/min)	FEED (mm/rev)	TOOLING
1	Hold sawn blank in chuck , leave 100 mm protruding .		–	–	3-jaw chuck
2	Set for true running with D T I and face end		320	hand	Facing tool & DTI
3	Centre drill and support job with revolving centre .		810	hand	Drill chuck, centre drill, revolving centre.
4	Turn 30 mm diameter for 90 mm length , leave shoulders square .		320	0·05	Bar turning tool
5	Turn 25 mm diameter for 51 mm length , leave shoulder square .		400	0·05	" " "
6	Turn 20 mm diameter for 27·5 mm length , leave shoulder square.		500	0·05	" " "
7	Drill through and ream 10 mm diameter .		{ 810 drill 144 ream } hand	{ ø9·5 drill ø10 reamer
8	Tap M12 x 1·75		Rotate by hand	hand	M12 x 1·75 tap Tap wrench.
9	Reverse in chuck and hold on 30 mm diameter .		–	–	3-jaw chuck
	rill, parallel for 64 mm .		500	hand	ø20 mm drill MT sleeve 2/3
				hand	Boring tool . Bed stop .

Fig. 2.16 *Typical training diary*

DAILY REPORT	
WEEK No. _Eleven_ FROM: _12 - 10 - 97_ TO: _17 - 10 - 97_	
MONDAY	Tech . day .
TUESDAY	Introduction to turning . We had a lecture / demonstration on the centre lathe. We were shown how to change a chuck and how to mount and use different tools. We then practiced facing , turning to a shoulder and drilling .
WEDNESDAY	Commenced turning exercise 10 . We were shown how to use a DTI and how to calculate and choose speeds and feeds. I turned all the external diameters at the first setting & drilled and reamed the hole (ø 10 mm) .
THURSDAY	Tapped the hole using a tap wrench to stop the tap turning and using the tailstock centre to guide the tap . The machine was isolated , put into neutral and the chuck was rotated by hand . Reversed work in chuck and finished ø45 mm . Drilled out to ø20 mm and bored ø30 mm to depth .
FRIDAY	Commenced work on next exercise (tap wrench) on the fitting section. Apprentice association meeting all afternoon .

2.7 Communications (miscellaneous)

Safety and hazard notices

There is a saying that *in an emergency people panic in their own language*. Therefore, all safety notices and operating instructions for potentially hazardous plant and processes should be printed in as many languages as there are employees from different ethnic backgrounds. Wherever possible, internationally recognised hazard signs should be used.

Safety and hazard signs

All signs must comply with the Safety Signs Regulations 1980. These are recognised internationally and combine geometrical shape, colour and a pictorial symbol to convey the message. Some examples are shown in Fig. 2.17.

Colour coding

This is another means of communications that overcomes language barriers. Table 2.1 shows the colour codes for the contents of gas cylinders. A cylinder which is coloured wholly red or maroon, or has a red band round it near the top, contains a flammable gas.

In the case of red cylinders, the name of the gas should also be stated on the cylinder. Maroon-coloured cylinders only contain acetylene gas for welding. A cylinder having a yellow band round the top contains a poisonous gas.

Fig. 2.17 *Safety hazard signs*

No smoking

Prohibition signs
Round with a white background and red border and cross bar. Black symbols placed centrally without obliterating the background. The sign means you must not ignore.

Caution, industrial trucks

Warning signs
Triangular with yellow background and a black border. Black symbols placed centrally. The sign warns of a particular hazard.

Foot protection must be worn

Mandatory signs
Round with a blue background and a white symbol. The sign states what protective equipment must be worn.

Indication of direction

Emergency signs
Square or oblong with white symbols on a green background. The sign indicates safe conditions such as emergency routes.

Table 2.1 *Colour codes for cylinder contents*

Gas	Ground colour of cylinder	Colour of bands
Acetylene	Maroon	None
Air	Grey	None
Ammonia	Black	Red and yellow
Argon	Blue	None
Carbon monoxide	Red (+ name)	Yellow
Coal gas	Red (+ name)	None
Helium	Medium brown	None
Hydrogen	Red (+ name)	None
Methane	Red (+ name)	None
Nitrogen	Dark grey	Black
Oxygen	Black	None

The identification colours for electric cables are shown in Table 2.2. Note that earthing conductors are likely to be referred to as *circuit protective conductors*.

Table 2.2 *Colour codes for electrical cables*

Service	Cable		Colour
Single phase	Live		Brown
Flexible	Neutral		Blue
	Earth		Green/yellow
Single phase	Live		Red
Non-flexible	Neutral		Black
	Earth		Green/yellow
Three phase		colour	Red
Non-flexible	Line (live) {	denotes	White
		phase	Blue
	Neutral		Black
	Earth		Green/yellow

Finally, pipe runs and conduit are colour coded according to their contents. The colours used are given in Table 2.3; the different possible method of application are shown in Fig. 2.18.

Table 2.3 *Colour codes for pipe contents*

Colour	Contents
White	Compressed air
Black	Drainage
Dark grey	Refrigeration and chemicals
Signal red	Fire (hydrant and sprinkler supplies)
Crimson or aluminium	Steam and central heating
French blue	Water
Georgian green	Sea, river and untreated water
Brilliant green	Cold water services from storage tanks
Light orange	Electricity
Eau-de-Nil	Domestic hot water
Light brown	Oil
Canary yellow	Gas

Fig. 2.18 *Colour codes for the contents of pipes*

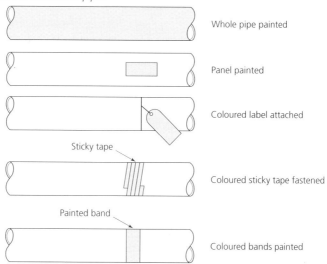

Whole pipe painted

Panel painted

Coloured label attached

Sticky tape

Coloured sticky tape fastened

Painted band

Coloured bands painted

Posters

Posters are also used to portray safety messages. They may be humorous or dramatic. The picture reinforces the caption so that the message is clear even for people who cannot read the words for some reason. Such posters should be displayed at strategic points adjacent to the hazard they represent. They should be changed frequently in order to attract attention. Examples are shown in Fig. 2.19.

Fig. 2.19 *Typical safety posters (source: RoSPA)*

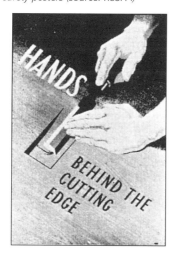

2.1 *Information required to undertake work tasks*
(a) You have just received the works order form for the next component you are to make. List the essential information you would expect to find on such a form.
(b) As well as the works order form, what additional and essential document do you need before you can start the task?

2.2 *Interpretation of numerical information*
(a) Graphs are often used for showing numerical relationships. Sketch suitable graphs to represent the following situations:
    (i) the relationship between the diameter and cross-sectional area of mild steel rods of 2, 4, 6, 8, 10 and 12 millimetres diameter
    (ii) the relationship between time and total power in watts for an office that has eight fluorescent electric lights; each light has a power rating of 80 watts; the lights are turned on, one at a time, at ten minute intervals until they are all on
    (iii) a firm has the following notable accident record:
        1994   15 accidents
        1995   24 accidents
        1996   12 accidents
        1997   7 accidents
(b) With the aid of simple examples, explain when you would use the following types of graphical representation:
    (i) ideograph (pictogram)
    (ii) pie chart

2.3 *Extraction and interpretation of engineering information*
(a) The parts list of a general arrangement drawing specifies the use of a manufacturer's standard drill bush with a bore of 6 mm and an o/d of 12 mm. State where you would look for details of such bushes and list the information you would need to give to the stores so that they could purchase such a bush.
(b) What do the following initial letters stand for: BSI, ISO, EN? (Note: you may find two uses of the initials EN.) To what do the specifications BS 970 and BS 308 refer?
(c) Explain briefly, with the aid of an example, what is meant by a *production schedule*.
(d) Give an example of a *product specification* for any product with which you are familiar.
(e) Table 2.4 shows an abstract from some screw thread tables. What is the pitch of an M10 thread and what tapping size drill is required for tapping an internal M10 screw thread?
(f) Figure 2.20 shows various types of engineering drawings and diagrams. Write down the letters of the individual illustrations and state the type of drawing shown.

Fig. 2.20 *Exercise 2.3(f)*

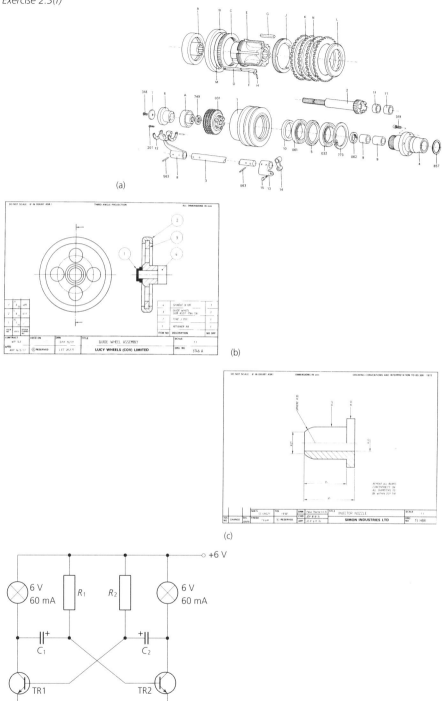

(a)

(b)

(c)

(d)

Table 2.4 *Exercise 2.3(e)*

150 metric threads (coarse series)	Minor dia. (mm)	Tensile stress area (mm²)	Tapping drill (mm)	ISO Hexagon (mm)
M0.8 × 0.2	0.608	0.31	0.68	—
M1.0 × 0.25	0.675	0.46	0.82	2.5
M1.2 × 0.25	0.875	0.73	1.0	3.0
M1.4 × 0.30	1.014	0.98	1.2	3.0
M1.6 × 0.35	1.151	1.27	1.35	3.2
M1.8 × 0.35	1.351	1.70	1.55	—
M2.0 × 0.40	1.490	2.1	1.70	4.0
M2.2 × 0.45	1.628	2.5	1.90	—
M2.5 × 0.45	1.928	3.4	2.20	5.0
M3.0 × 0.5	2.367	5.0	2.65	5.5
M3.5 × 0.6	2.743	6.8	3.10	—
M4.0 × 0.7	3.120	8.8	3.50	7.0
M4.5 × 0.75	3.558	11.5	4.0	—
M5.0 × 0.8	3.995	14.2	4.50	8.0
M6.0 × 1.0	4.747	20.1	5.3	10.0
M8.0 × 1.25	6.438	36.6	7.1	13.0
M10.0 × 1.50	8.128	58.0	8.8	17.0
M12.0 × 1.75	9.819	84.3	10.70	19.0
M16.0 × 2.00	13.510	157.0	14.5	24.0

2.4 *Evaluation of the accuracy and appropriateness of engineering information*

(a) Give **two** reasons for cross-checking the accuracy of any reference books that might be lying around in your workshop.

(b) To whom should you refer for guidance on the accuracy and relevance of reference material available in your workshop?

(c) Should you believe advice given to you by experienced members of the workforce, or should you check the validity of such advice, and, if so, why?

2.5 *Recording and processing engineering information*

(a) State **four** reasons for, and the importance of, accurate record keeping in a modern factory environment.

(b) State whether it is a legal requirement to keep a log of notifiable accidents and, if so, the person who has the authority to demand access to such a log.

(c) Describe briefly how quality control is maintained in your company, or your training centre, and what records are required.

2.6 *Methods of record keeping*

(a) Computer files have superseded many manual filing systems. Why should back-up copies of files be kept, and how can these be kept?

(b) For what purposes are the following methods of record keeping used?
   (i) log books (other than your training log book)
   (ii) forms and schedules
   (iii) photographic (pictorial and dye-line)
   (iv) drawings and diagrams

(c) Why is it important to keep a training log book, and why should it be kept carefully, away from dirt and oil, so that it is always clean, neat and tidy?

3 Engineering drawing

When you have read this chapter, you should understand how to:

- Interpret (read) drawings in first- and third-angle projection.
- Sketch and dimension mechanical components in first- and third-angle projection.
- Sketch mechanical components in isometric and oblique projection.

3.1 Engineering drawing

Figure 3.1(a) shows a drawing of a simple clamp. This is a pictorial drawing. It is very easy to see what has been drawn, even to people who have not been taught how to read an engineering drawing. Unfortunately such drawings have only a limited use in engineering. If, onto this drawing, you try to put all the information required to make the clamp, it would become very cluttered and difficult to interpret. Therefore we use a system called *orthographic drawing* when we make engineering drawings.

An example of an orthographic drawing of our clamp is shown Fig. 3.1(b). We now have a collection of drawings, each one looking at the clamp from a different direction. This enables us to show every feature of the clamp that can be seen, and also some things that cannot be seen (hidden details). Things that cannot be seen are indicated by broken lines. Finally we can add the sizes (dimensions) that we need in order to make the clamp. These are shown in Fig. 3.1(c).

You do not have to know how to produce a formal engineering drawing using a drawing board and T-square, or a draughting machine, or a CAD package on a computer, but you must be able to make sketches and you must be able to interpret formally produced orthographic drawings sent down to the machine shop from the drawing office. In case you are wondering, CAD stands for Computer Aided

Design/Draughting. A drawing that has all the information required to make a component part, such as Fig. 3.1(c), is called a *detail drawing*, but more of that later.

Fig. 3.1 *Clamp: (a) pictorial drawing; (b) orthographic drawing; (c) fully dimensioned. (Dimensions in millimetres)*

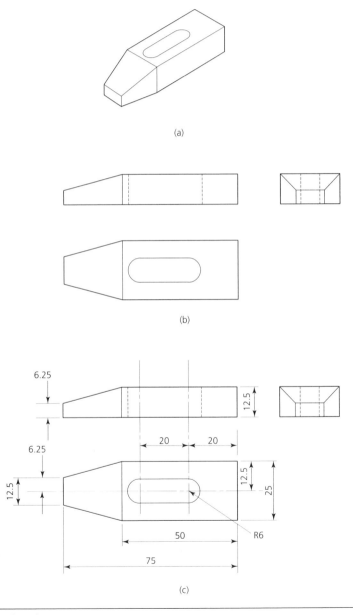

3.2 First-angle orthographic projection

There are two systems of orthographic drawing used by engineers:

- First-angle or English projection.
- Third-angle or American projection.

In this section we are going to look at *first-angle* projection. We are again going to use the clamp you first met in Fig. 3.1(a) because it is nice and simple. We look at the clamp from various directions:

- We can look down on the top of the clamp and draw what we see. This is called a *plan view*. What we see is shown as view **1** in Fig. 3.2(a).

Fig. 3.2 *Principles of drawing in first-angle projection: (a) plan views, end view, and elevation (side view); (b) collected views together make up an orthographic drawing*

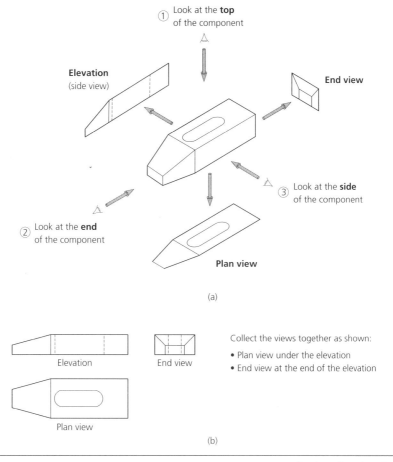

- We can look at the end of the clamp and draw what we see. This is called an *end view*. What we see is shown in as view **2** Fig. 3.2(a).
- We can look at the side of the clamp and draw what we see. Although this is a side view, it is given a special name. It is called an *elevation*. What we see is shown as view **3** in Fig. 3.2(a).
- We can now assemble these views together in the correct order, as shown in Fig. 3.2(b). We now have a *first-angle orthographic drawing* of the clamp.

As well as the things we can see from the outside of the clamp, we also included some 'hidden detail' in the end view and elevation. Why did we do this? If we had only shown the slot as an oval in the plan view it could have meant one of two things:

- A slot passing right through the clamp.
- A slot recessed part way into the clamp.

It could not have been an oval-shaped lump on top of the clamp as this would have shown up in the end view and in the elevation. Figure 3.3(a) shows a slot going right through the clamp and Fig. 3.3(b) shows a slot going only part way through the clamp (a pocket). Figure 3.3(c) shows how the elevation would have appeared if the oval had been a projection on top of the clamp. Hidden detail is shown by a thin dashed line. Figure 3.3(d) shows how the plan view could have been common to all the above possibilities.

Fig. 3.3 *Hidden detail: (a) slot passes right through the clamp; (b) slot passes part way through the clamp; (c) no hidden details, therefore no slot, therefore oval can only be a projection as shown; (d) the plan view is the same for (a), (b) and (c) – the elevation is required to show what the oval refers to*

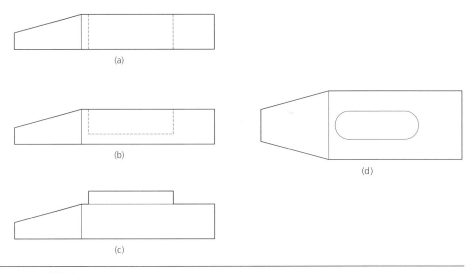

So let us summarise how we draw in *first-angle projection* and how we know a drawing is in first-angle projection.

- To draw the plan view we look down on top of the component and draw the plan view underneath the side view or elevation.
- To draw the end view we look at the end of the component and draw the end view at the opposite end of the component to the one we are looking at.
- The end view can be drawn at either end or at both, as long as the rules are obeyed.
- Sometimes only two views are used when the plan and elevation are the same. For example, Fig. 3.4 shows that an elevation and an end view provides all the information we require for a cylindrical component such as a shaft.

Fig. 3.4 *First-angle drawing of a cylindrical component: the elevation and plan views are the same and need only be drawn once*

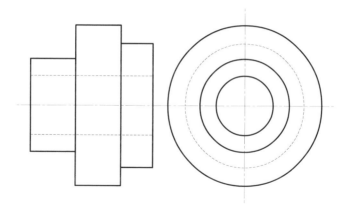

Finally let us see how an orthographic drawing is constructed.

- First we draw the ground lines and a plane at 45°, as shown in Fig. 3.5(a).
- We then start to draw in the construction lines faintly, using lines that are half the thickness of the final outline. Figure 3.5(b) shows the construction lines in place.
- We should follow each construction line round all the views where it appears to avoid confusion.
- Finally, we 'line in' the outline so that it stands out boldly, as shown in Fig. 3.5(c).

Fig. 3.5 *Making a drawing in first-angle projection: (a) ground line and planes; (b) initial construction lines; (c) line in the outline (outline is twice the thickness of the construction lines)*

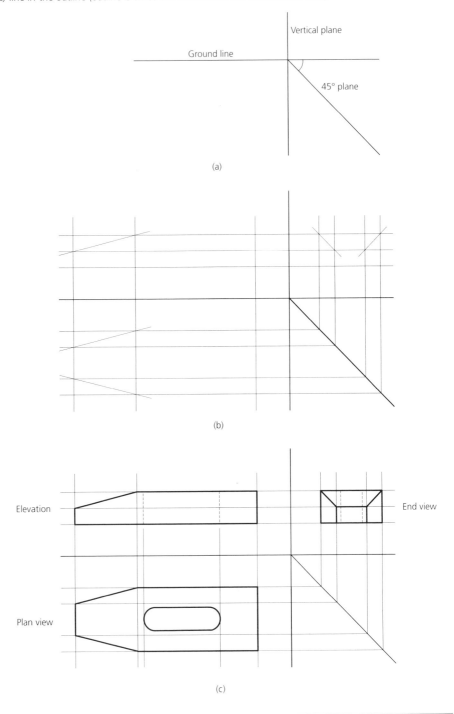

Vertical plane

Ground line

45° plane

(a)

(b)

Elevation

End view

Plan view

(c)

3.3 Third-angle orthographic projection

To draw our clamp in third-angle orthographic projection, we merely have to rearrange the relative positions of the views. Each view now appears at the same side or end of the component from which we are looking at it. This is shown in Fig. 3.6. That is:

- We look down on the clamp and draw the plan view above the side view or elevation.
- We look at the left-hand end of the clamp and draw the end view at the same end.

Fig. 3.6 *Principles of drawing in third-angle projection*

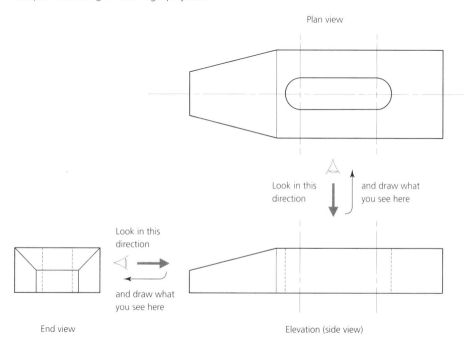

Plan view

Look in this direction | and draw what you see here

Look in this direction

and draw what you see here

End view

Elevation (side view)

So what is the advantage of third-angle projection? Let us consider the general arrangement drawing for an airliner drawn to a fairly large scale so that fine detail can be shown. In first-angle projection, the end view, looking at the nose of the aircraft, would be drawn somewhere beyond the tail. An end view looking at the tail of the aircraft would be drawn somewhere beyond the nose. It is much more convenient to draw the end view of the nose of the aircraft at the nose end of the elevation. Also, it is more convenient to draw an end view of the tail of an aircraft next to the tail of the elevation.

With experience, we can determine which projection is being used by the arrangement of the views. Nevertheless, to avoid confusion, we always state the projection used on the drawing. Sometimes the projection used is stated in words, but more usually it is indicated by the use of a standard symbol. Figure 3.7(a) shows the standard symbol for first-angle projection and Fig. 3.7(b) shows some examples of simple shapes drawn in first-angle projection.

Fig. 3.7 *First-angle orthographic projection: (a) symbol; (b) examples of shapes drawn in first-angle projection*

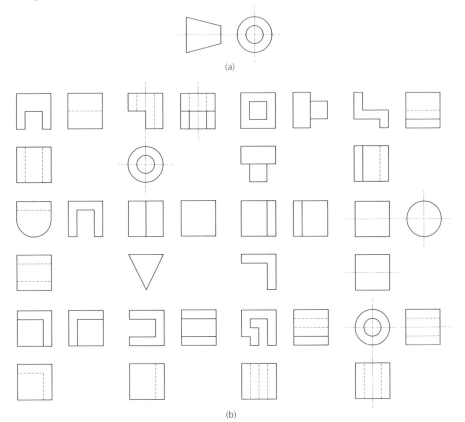

Figure 3.8(a) shows the standard symbol for third-angle projection and Fig. 3.8(b) shows some examples of simple shapes drawn in third-angle projection. I hope you can see the difference by comparing Figs 3.7 and 3.8. Figure 3.8(c) shows the combined first- and third-angle symbol and how it is used.

So far we have only considered features that are conveniently arranged at right angles to each other so that their true shape is shown in the plan, elevation or the end view. This is not always the case and sometimes we have to include an *auxiliary view*. This technique is important in the production of working drawings to enable the positions of features on an inclined surface to be dimensioned. Figure 3.9 shows a bracket with an inclined face. When it is drawn in first-angle projection, it can be seen that the end view showing the inclined surface and its features are heavily distorted. However, they appear correct in size and in shape in the auxiliary view (AV), which is projected at right angles (perpendicular) to the inclined face.

Fig. 3.8 *Third-angle orthographic projection; (a) symbol; (b) examples of shapes drawn in third-angle projection; (c) use of the combination symbol*

(a)

(b)

Combination symbol printed on drawing sheets where the system of projection is optional

Deleting a symbol to indicate first angle orthographic projection

Deleting a symbol to indicate third angle orthographic projection

(c)

Fig. 3.9 *Auxiliary view: EL = elevation; EV = end view; PL = plan; AV = auxiliary view*

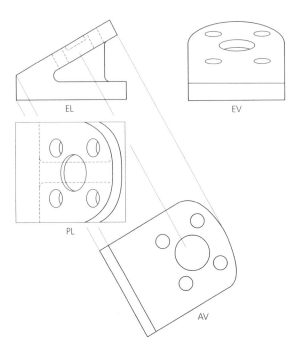

EL

EV

PL

AV

3.4 Conventions

An engineering drawing is only a means of recording the intentions of the designer and communicating those intentions to the manufacturer. It is not a work of art and, apart from the time spent in its preparation, it has no intrinsic value. If a better and cheaper method of communication could be discovered, then the engineering drawing would no longer be used. We are already part way along this road with CAD where the drawings are stored digitally on magnetic or optical disks and can be transmitted between companies by the Internet. However, hard copy, in the form of a printed drawing, still has to be produced for the craftsperson or the technician to work to.

As an aid to producing engineering sketches and drawings quickly and cheaply, we use *standard conventions*. These are recognised internationally and are used as a form of drawing 'shorthand' for the more frequently used and the more awkward details. For example, you can see from Fig. 3.10 that it is easier to represent a screw thread by the standard convention than it is to draw out the thread pictorially.

In the UK we use the *British Standard for Engineering Drawing Practice* as published by the British Standards Institute (BSI). This standard is based upon the recommendations of the International Standards Organisation (ISO) and, therefore, its conventions and guidelines, and drawings produced using such conventions and guidelines, are accepted internationally. The standard is published in three parts:

- BS 308: Part 1: 1984: *Recommendations for general principles.*
- BS 308: Part 2: 1985: *Recommendations for dimensioning and tolerancing of size.*
- BS 308: Part 3: 1990: *Recommendations for general tolerancing.*

Fig. 3.10 *Need for conventions: (a) BS convention; (b) pictorial representation*

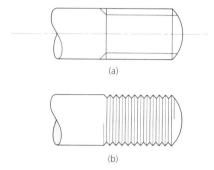

(a)

(b)

3.4.1 *Types of line*

Table 3.1 and Figure 3.11 show the types of line recommended by BS 308, together with some typical applications. The following points should be noted in the use of these lines.

Table 3.1 *Abbreviations for written statements*

Term	Abbreviation	Term	Abbreviation
Across flats	A/F	Hexagon head	HEX HD
British Standard	BS	Material	MATL
Centres	CRS	Number	NO.
Centre line	CL or ℄	Pitch circle diameter	PCD
Chamfered	CHAM	Radius (in a note)	RAD
Cheese head	CH HD	Radius (preceding	
Countersunk	CSK	a dimension)	R
Countersunk head	CSK HD	Screwed	SCR
Counterbore	C'BORE	Specification	SPEC
Diameter (in a note)	DIA	Spherical diameter	
Diameter (preceding		or radius	SPHERE ∅ or R
a dimension)	∅	Spotface	S'FACE
Drawing	DRG	Standard	STD
Figure	FIG.	Undercut	U'CUT
Hexagon	HEX		

Fig. 3.11 *Types of line and their applications*

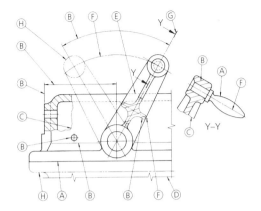

Line		Description	Application
A	————	Continuous thick	Visible outlines and edges
B	————	Continuous thin	Dimension, projection and leader lines, hatching, outlines of revolved sections, short centre lines, imaginary intersections
C	⌇	Continuous thin irregular	Limits of partial or interrupted views and sections, if the limit is not an axis
D	─╲─╲─	Continuous thin straight with zigzags	
E	----------	Dashed thin	Hidden outlines and edges
F	— — — —	Chain thin	Centre lines, lines of symmetry, trajectories and loci, pitch lines and pitch circles
G	⌐·— / —·⌐	Chain thin, thick at ends and changes of direction	Cutting planes
H	— ·· — ·· —	Chain thin double	Outlines and edges of adjacent parts, outlines and edges of alternative and extreme positions of movable parts, initial outlines prior to forming, bend lines on developed blanks or patterns

- *Dashed lines* These should consist of dashes of consistent length and spacing, approximately to the proportions shown in the figure.
- *Thin chain lines* These should consist of long dashes alternating with short dashes. The proportions should be generally as shown in the figure, but the lengths and spacing may be increased for very long lines.
- *Thick chain lines* These should have similar lengths and spacing as for thin chain lines.
- *General* All chain lines should start and finish with a long dash. When thin chain lines are used as centre lines, they should cross one another at solid portions of the line. Centre lines should extend only a short distance beyond the feature unless required for dimensioning or other purposes. They should not extend through the spaces between the views and should not terminate at another line of the drawing. Where angles are formed in chain lines, long dashes should meet at the corners and should be thickened, as shown. Arcs should join at tangent points. Dashed lines should also meet at corners and tangent points with dashes.

3.4.2 *Conventions*

Figure 3.12 shows some typical conventions used in engineering drawings. It is not possible, in the scope of this book, to provide the full set of conventions or to provide detailed explanations of their use. For this it is necessary to consult texts specialising in engineering drawing together with British Standard 308. The full standard is expensive but you should find the special abridged edition, BS PP7308: 1986: *Engineering drawing practice for schools and colleges*, adequate for your needs. This edition is published at a very affordable price.

Fig. 3.12 *Typical conventions for some common features*

Title	Subject	Convention	
External screw threads (detail)			
Internal screw threads (detail)			
Diamond knurling			
Square on shaft			
Holes on circular pitch			
Bearings			
Title	**Subject**	**Convention**	**Diagrammatic representation**
Cylindrical compression spring	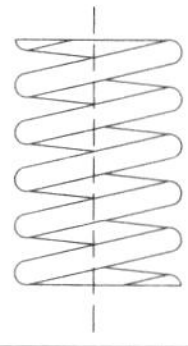	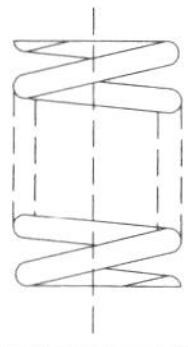	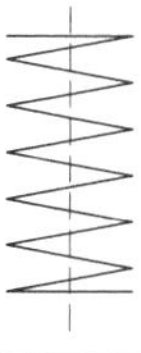

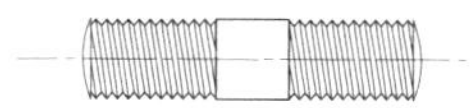

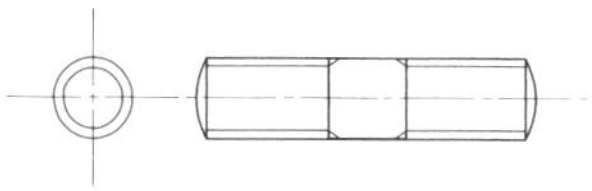

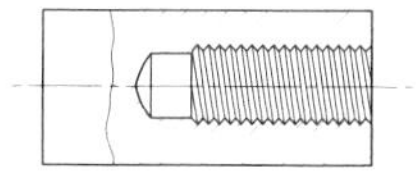

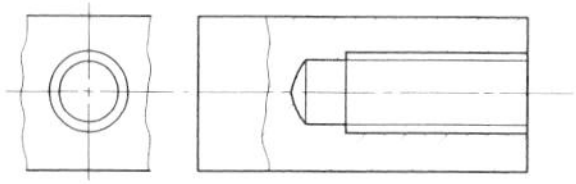

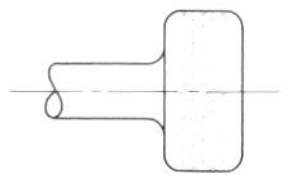

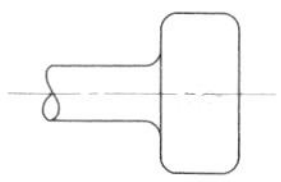

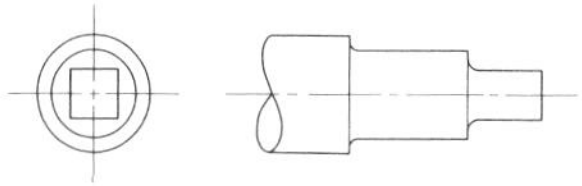

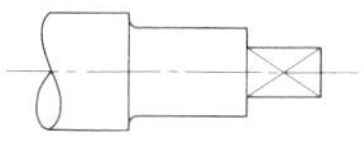

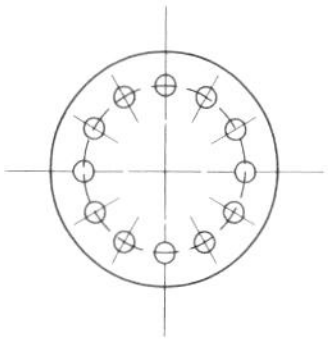

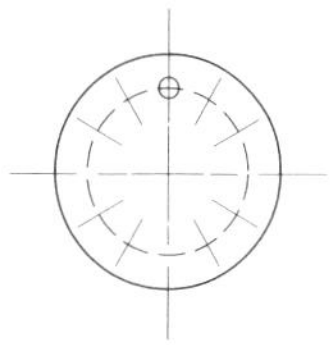

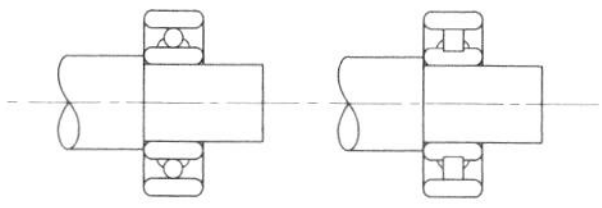

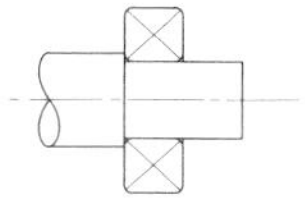

3.4.3 *Abbreviations for written statements*

Table 3.1 lists the standard abbreviations for written statements as used on engineering drawings. Some examples of their use are shown in Fig. 3.13. Some further examples will be given when we discuss the dimensioning of drawings. Figure 3.13(a) shows the abbreviation and 3.13(b) shows the corresponding interpretation

Fig. 3.13 *Examples of the use of standard abbreviations: (a) counterbored hole; (b) countersunk hole*

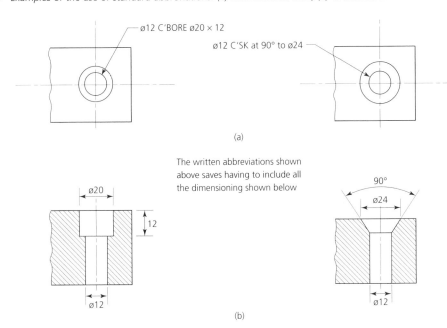

3.5 Redundant views

It has been stated and shown earlier that where a component is symmetrical you do not always require all the views to provide the information required for manufacture. A ball looks the same from all directions, and to represent it by three circles arranged as a plan view, an elevation and an end view would just be a waste of time. One circle and a note that the component is spherical is all that is required. The views that can be discarded without loss of information are called *redundant views*. Figure 3.14 shows how drawing time can be saved and the drawing simplified by eliminating the redundant views when drawing symmetrical components.

Fig. 3.14 *Redundant views: (a) first-angle working drawing of a symmetrical component (plan view redundant); (b) symmetrical component reduced to two views; (c) working drawing reduced to single view by using revolved sections and BS convention for the square flange*

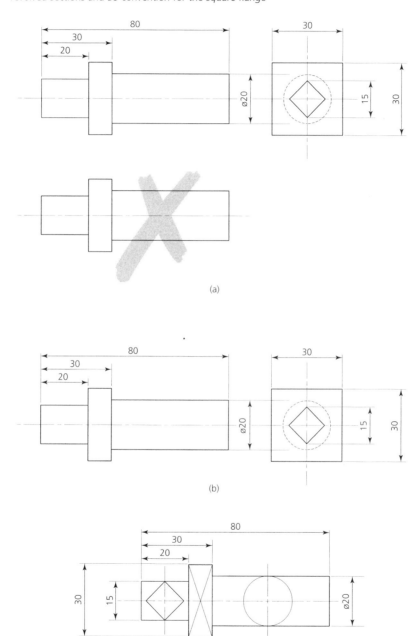

3.6 Dimensioning

So far, only the shape of the component has been considered. However, in order that components can be manufactured, the drawing must also show the size of the component and the position and size of any features on the component. To avoid confusion and the chance of misinterpretation, the dimensions must be added to the drawing in the manner laid down in BS 308. Figure 3.15(a) shows how projection and dimension lines are used to relate the dimension to the drawing, whilst Fig. 3.15(b) shows the correct methods of dimensioning a drawing.

Fig. 3.15 *Dimensioning: (a) projection and dimension lines; (b) correct and incorrect dimensioning*

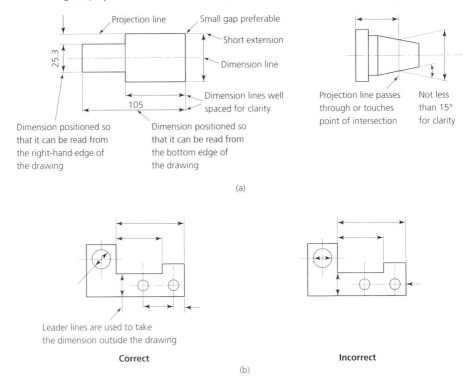

3.6.1 *Correct dimensioning*

- Dimension lines should be thin full lines not more than half the thickness of the component outline.
- Wherever possible, dimension lines should be placed outside the outline of the drawing.
- The dimension line arrowhead must touch, but not cross, the projection line.
- Dimension lines should be well spaced to ensure that the numerical value of the dimension can be clearly read and that they do not obscure the outline of the drawing.

3.6.2 *Incorrect dimensioning*

- Centre lines and extension lines must **not** be used as dimension lines.
- Wherever possible, dimension line arrowheads must not touch the outline directly but should touch the projection lines which extend from the outline.
- If the use of a dimension line within the outline is unavoidable, then try to use a leader line to take the dimension itself outside the outline.

Figure 3.16(a) shows how circles and shaft ends (circles) should be dimensioned. It is preferable to use those techniques which take the dimension outside the circle, unless the circle is so large that the dimension will neither be cramped nor obscure some vital feature. Note the use of the symbol ⌀ to denote a diameter. Figure 3.16(b) shows how radii should be dimensioned. Note that the radii of arcs of circles need not have their centres located if the start and finish points are known. Figure 3.16(c) shows how notes may be used to avoid the need for the full dimensioning of certain features of a drawing.

Fig. 3.16 *Dimensioning – diameters and radii: (a) dimensioning holes; (b) dimensioning the radii of arcs which need not have their centres located; (c) use of notes to save full dimensioning*

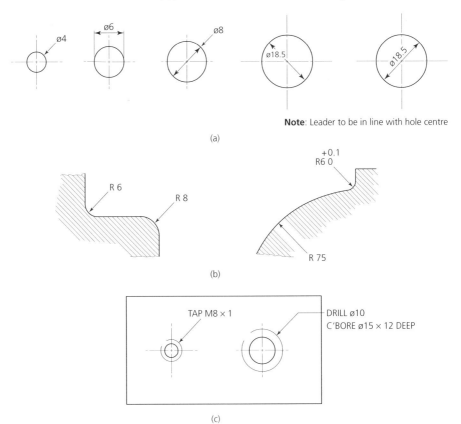

Figure 3.17 shows how leader lines indicate where notes or dimensions are intended to apply. Such lines are terminated with arrowheads or dots. *Arrowheads* are used where the leader line touches the outline of a component or feature. *Dots* are used where the leader line finishes within the outline of the component or feature to which it refers.

In the same way that the number of views in a drawing are kept to a minimum, dimensions are selected to avoid confusion by unnecessary detail. Duplication of dimensions must also be avoided. The dimensions should appear adjacent to the view in which the feature being dimensioned appears as a true shape and size. For example, a hole should be dimensioned in the view in which it appears as a circle and not where it appears as hidden detail. Figure 3.18 shows how dimensions should be correctly positioned.

Fig. 3.17 *Termination of leader lines: (a) leader lines terminating in arrowheads; (b) leader lines terminating in dots*

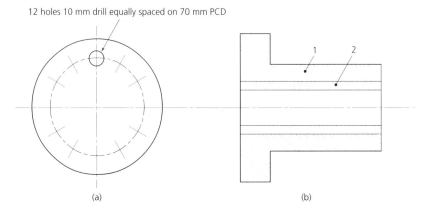

Fig. 3.18 *Correct positioning of dimensions: dimensions should be placed in the view which shows the feature in clear outline. (Dimensions in millimetres)*

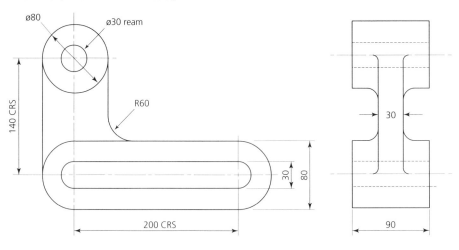

3.6.3 *Auxiliary dimensions*

It has already been stated that, to avoid mistakes, duplicated or unnecessary dimensions should not appear on a drawing. The only exception to this rule is when *auxiliary dimensions* are used to avoid the calculation of, say, overall dimensions on the shop floor. Such auxiliary dimensions are placed in brackets as shown in Fig. 3.19. The dimensions marked *F* are the *functional* dimensions from which the component is manufactured. Auxiliary dimensions are also sometimes referred to as *non-functional* dimensions.

Fig. 3.19 *Auxiliary dimensions: (a) overall size added as auxiliary dimension; (b) hole centres added as auxiliary dimension. (F = functional dimension; dimensions in millimetres)*

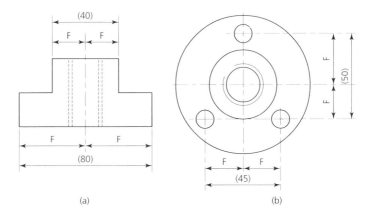

(a)　　　　　　　　　　　　　　　　　(b)

3.7 Toleranced dimensions

It is true to say if ever a component was made exactly to size no one would ever know because it could not be measured exactly. Having calculated the ideal size for a dimension, the designer must then decide how much variation from that size he will tolerate. This variation between the smallest and the largest acceptable size is called the *tolerance*. As well as specifying the magnitude of the tolerance the designer must also indicate where the tolerance lies relative to the nominal size. Figure 3.20(a) shows various methods of tolerancing a dimension and how the tolerance is interpreted. Figure 3.20(b) shows the types of 'fit' that can be achieved between mating components. Figure 3.20(c) explains the terms used.

Fig. 3.20 *Toleranced dimensions: (a) methods of tolerancing; (b) classes of fit; (c) limit system definitions. (Dimensions in millimetres)*

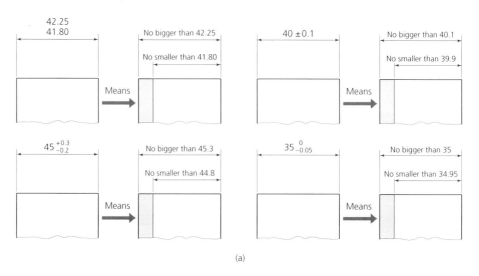

(a)

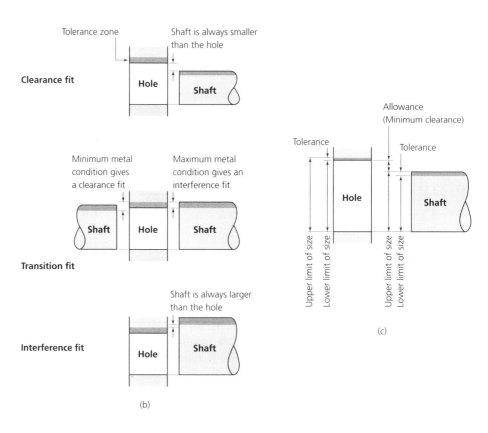

(b)

(c)

When toleranced dimensions are used, cumulative errors can occur wherever a feature is controlled by more than one toleranced dimension, as shown in Fig. 3.21(a). It can be seen that chain (string) dimensioning gives a build-up of tolerances which is greater than the designer intended. In this example the maximum tolerance for the right-hand hole centre is three time the individual tolerances. That is, the sum of the individual tolerances is $(\pm 0.1) + (\pm 0.1) + (\pm 0.1)$ mm $= \pm 0.3$ mm from the left-hand datum edge. This *cumulative* effect can be easily eliminated by dimensioning each feature individually from a common datum, as shown in Fig. 3.21(b).

Fig. 3.21 *Cumulative error: (a) string dimensioning – cumulative tolerance equals sum of individual tolerances; (b) dimensioning from one common datum to eliminate cumulative effect. (Dimensions in millimetres)*

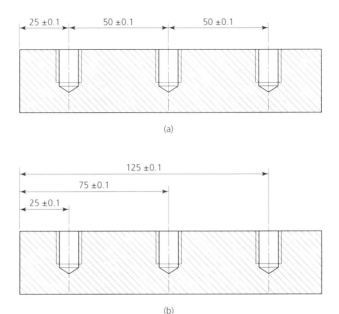

Cumulative errors can also occur when a number of components are assembled together, as shown in Fig. 3.22. To avoid this, assembly dimensions are given. The individual components are adjusted within their individual tolerances, or selective assembly is used so that the cumulative error is within the tolerance of the assembly dimension. In the example shown, the dimension A of the assembled components can vary as follows for the component tolerances given:

Maximum $12.08 + 75.05 + 12.08 = 99.21\,mm$

Minimum $11.92 + 74.95 + 11.92 = 98.79\,mm$

This variation may be too great for the assembled components to function properly. Thus an assembly dimension should be given specifying the dimension A. In this example the dimension $A = 99 \pm 0.01$ mm. This assembled dimension is achieved by selective assembly, by adjusting the sizes of the parts within their individual tolerances or by correction after assembly.

Fig. 3.22 *Assembly dimensions. (Dimensions in millimetres)*

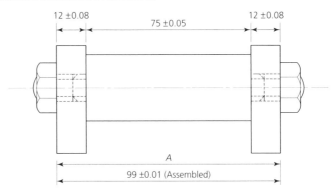

It is not usually necessary to tolerance every individual dimension, only the important ones. The rest can be given a *general tolerance* in the form of a note in the title block, as shown in Fig. 3.23. This general statement may refer either to *open dimensions* or it may say *except where otherwise stated*. The general tolerance is 0.5 mm with the limits stated as +0.3 mm and −0.2 mm in the example shown in Fig. 3.23(a) and with limits stated as ±0.2 mm in the example shown in Fig. 3.23(b). Both examples mean the same thing. Applied to an open dimension of 12 mm, the actual size is acceptable if it lies as shown in Fig. 3.23(c).

Fig. 3.23 *General tolerances*

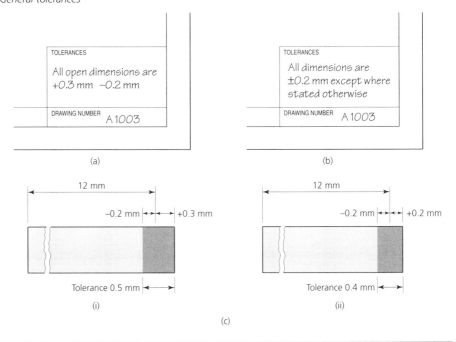

3.8 Sectioning

Sectioning is used to show the internal details of engineering components which cannot clearly be shown by other means. The stages of making a sectioned drawing are shown in Fig. 3.24. It should be realised that steps (a), (b) and (c) are performed mentally in practice and only (d) is actually drawn.

Fig. 3.24 *Section drawing: (a) the clamp is to be sectioned along the line X–X; (b) the cutting plane is positioned on the line X–X as shown; (c) that part of the clamp that lies in front of the cutting plane is removed leaving the sectioned component; (d) sectioned orthographic elevation of the clamp shown in (a) – note that section shading lines lie at 45° to the horizontal and are half the thickness of the outline*

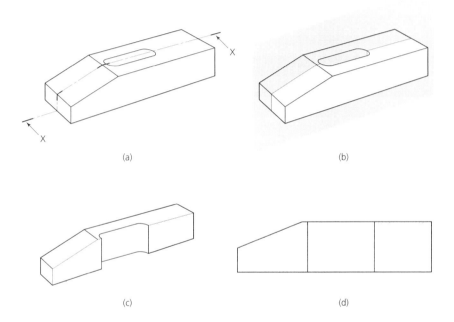

(a)   (b)

(c)   (d)

The rules for producing and reading sectioned drawings can be summarised as follows:

- Drawings are only sectioned when it is impossible to show the internal details of a component in any other way.
- Bolts, studs, nuts, screws, keys, cotters and shafts are not usually shown in section even when the cutting plane passes through them.
- Ribs and webs are not sectioned when parallel to the cutting plane.
- The cutting plane must be indicated in the appropriate view.
- Hidden detail is not shown in sectioned views when it is already shown in another view.
- The section shading (hatching) is normally drawn at 45° to the outline of the drawing using thin, continuous lines that are half the thickness of the outline. If the outline contains an angle of 45° then the hatching angle can be changed to avoid confusion.

- Adjacent parts are hatched in opposite directions. To show more than two adjacent parts, the spacing between the hatched lines can be varied. Some practical examples of sectioning are shown in Figs 3.25 and 3.26.

Fig. 3.25 *Further sections*

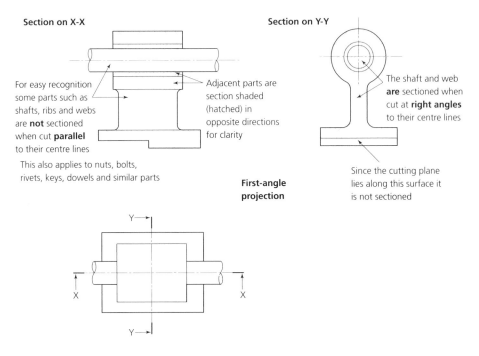

Section on X-X

Section on Y-Y

For easy recognition some parts such as shafts, ribs and webs are **not** sectioned when cut **parallel** to their centre lines

This also applies to nuts, bolts, rivets, keys, dowels and similar parts

Adjacent parts are section shaded (hatched) in opposite directions for clarity

First-angle projection

The shaft and web **are** sectioned when cut at **right angles** to their centre lines

Since the cutting plane lies along this surface it is not sectioned

3.9 Machining symbols

Machining symbols and instructions are used to:

- Specify a particular surface finish.
- Determine a machining process.
- Define which surfaces are to be machined.

Figure 3.27(a) shows the standard machining symbol (BS 308) and the proportions, in millimetres, to which it should be drawn. When applied to views of a drawing, as shown in Fig. 3.27(b), the symbol should be drawn as follows:

- Normal to a surface.
- Normal to a projection line.
- Normal to an extension line.
- As a general note.
  ('*Normal*' means '*at right angles to*' in this context.)

Fig. 3.26 *Miscellaneous sections: (a) sectioning in two planes; (b) the half section – this can be used with symmetrical components to show internal and external detail in the same view; the external view does not include hidden detail unless this is required for clarity or for dimensioning purposes; (c) revolved sections – these are a superimposed cross-section drawn on an outline view; where the section interferes with the outline, the outline is broken; (d) removed section; (e) scrap section*

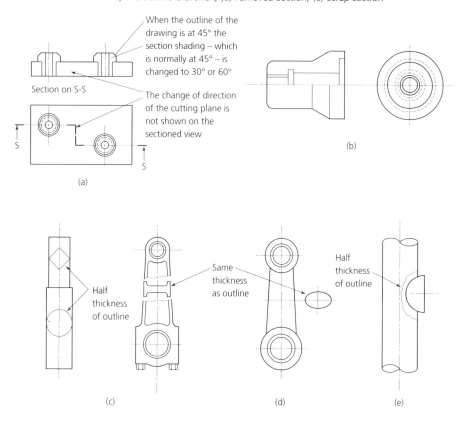

Because a machining symbol is interpreted as a precise instruction, its form should be drawn carefully. Figure 3.27(c) shows three fundamental variations of the symbol which, in turn, instruct the craftsperson to:

- Machine.
- Machine if necessary.
- Do *not* machine.

These symbols must be used carefully; one incorrect symbol or incorrect application of a symbol can result in unnecessary manufacturing costs or even the scrapping of a component. The symbol can be used as shown in Fig. 3.27(d) to indicate the quality of surface finish (acceptable roughness) and/or the production process to be used. Note that the unit of measurement applied to surface finish (roughness) is the micrometre (μm), that is, 0.001 mm. The surface finish value is printed above the machining symbol and, for very precise components, it is more usual to specify upper and lower limits of roughness values as shown.

Fig. 3.27 *The machining symbol: (a) drawing a machining symbol; (b) applying the machining symbol; (c) the machining symbol as an instruction; (d) specifying surfaces texture on a casting – dimensions omitted for clarity*

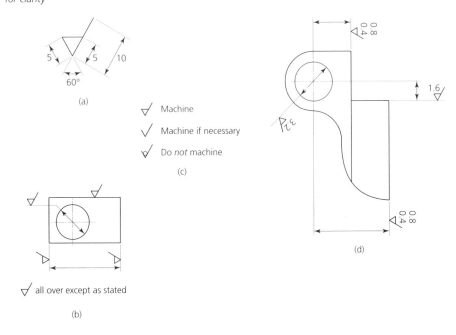

The surface finish (roughness value) can be related to manufacturing processes, as shown in Fig. 3.28. The roughness value itself represents the average roughness of the surface. Even a surface that appears smooth to the unaided eye would, if magnified, reveal thousands of minute hills and valleys.

Production processes related to a given surface finish can be specified by stating the process to be used on an extension of the symbol. Figure 3.29(a) shows how the upper and lower limits of roughness values and the machining process can be applied to the machining symbol. Figure 3.29(b) shows how a component can be completely dimensioned with both tolerances and machining symbols.

3.10 Engineering drawings

As has already been stated, craftspersons are not called upon to produce formal technical drawings, although they often have to make sketches of parts for maintenance replacements. However, a detailed knowledge of the types of drawing, the projection system used and the standard conventions is essential for the correct interpretation of drawings in order to manufacture the parts shown and to assemble them correctly. You have already read about the first- and third-angle projection systems, conventions, dimensions and other features earlier in this chapter, it is now time to consider the different types of drawing you will meet and how they are presented.

Fig. 3.28 *Roughness values. (Note: The ranges shown above are typical of the processes listed; higher or lower values may be obtained under special conditions)*

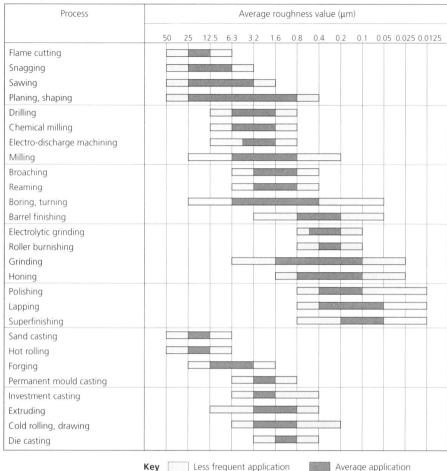

Process	Average roughness value (µm)
	50 25 12.5 6.3 3.2 1.6 0.8 0.4 0.2 0.1 0.05 0.025 0.0125
Flame cutting	
Snagging	
Sawing	
Planing, shaping	
Drilling	
Chemical milling	
Electro-discharge machining	
Milling	
Broaching	
Reaming	
Boring, turning	
Barrel finishing	
Electrolytic grinding	
Roller burnishing	
Grinding	
Honing	
Polishing	
Lapping	
Superfinishing	
Sand casting	
Hot rolling	
Forging	
Permanent mould casting	
Investment casting	
Extruding	
Cold rolling, drawing	
Die casting	

Key ☐ Less frequent application ▨ Average application

Fig. 3.29 *Complete dimensioning: (a) process specification; (b) application of limits and machining symbols*

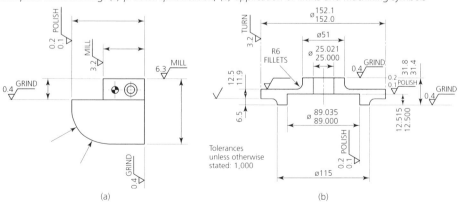

(a)   (b)

Engineering drawing **71**

To save time in the drawing office, most companies adopt a standardised and pre-printed drawing sheet, as shown in Fig. 3.30. The layout and content will vary from company to company but, generally, such sheets will provide the following information:

- The drawing number and title.
- The projection used (first angle or third angle).
- The scale.
- The general tolerance.
- The material specification.
- Warning notes.
- Any corrections or revisions, the date these were made, and the zone in which they occur.
- Special notes concerning, for example, heat-treated, decorative, corrosion-resistant or other surface finishes.

If manual drawing is to be used then a pre-printed tracing sheet on tracing paper or plastic sheet will be used. The latter is more expensive but it is more durable if many copies of the drawing are required over an extended period of time. Modern drawing offices are increasingly using CAD systems. The standard layout is saved in the memory of the computer as a 'template' and can be called up by a key-stroke when a drawing is to be made.

Fig. 3.30 *Layout of drawing sheet*

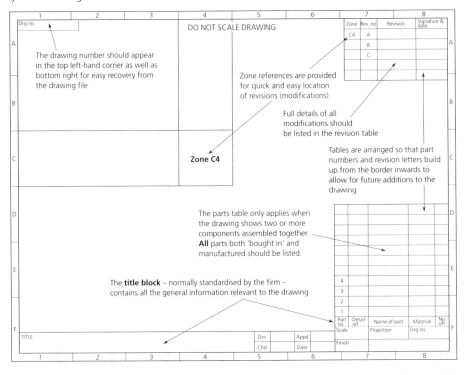

3.10.1 *Types of drawing*

General arrangement drawings

An example of a general arrangement (GA) drawing is shown in Fig. 3.31. It shows all the components correctly assembled, and lists all the parts required. For those parts that will be 'bought-in', it will state the maker and catalogue reference for the benefit of the purchasing department. For those parts to be made in the factory, the detail drawing numbers will be provided together with the material specification and the quantity of parts required. General arrangement drawings do not normally carry dimensions except, occasionally, overall dimensions for reference only.

Fig. 3.31 *General arrangement drawing*

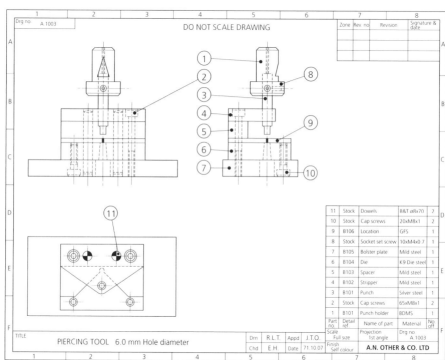

Detail drawings

These provide all the details required to make a component, and an example is shown in Fig. 3.32. It not only shows the shape of the component but also its size and the tolerances within which it must be manufactured. The detail drawing states the materials to be used and the heat treatment of the punch. Similar detail drawings would be required for all the other components shown in the general arrangement drawing.

Fig. 3.32 *Detail drawing*

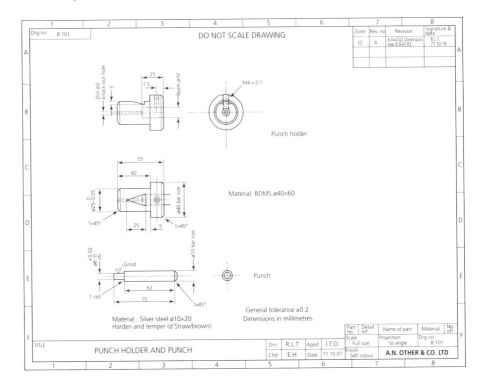

Exploded (assembly) drawings

You will find these mainly in the service manuals for machines and similar devices. They show the components in the correct relationship to each other and the stock reference number for each component to facilitate the ordering of spare parts. An example is shown in Fig. 3.33.

Block diagrams

You use these to show the relationship between the main subassemblies of machines. Figure 3.34(a) shows the drive train for a milling machine and Fig. 3.34(b) shows the drive train for a centre lathe.

Location (installation) drawings

These are used to show the correct layout of the various major items of a plant installation. It may also include details of any foundations which may have to be prepared and any service ducts which may have to be provided for electrical, hydraulic, pneumatic and fuel links and services. Two examples are shown in Fig. 3.35.

Fig. 3.33 *Exploded view and parts list (source: Myford Ltd)*

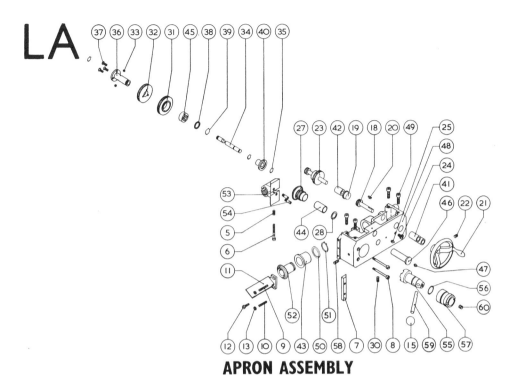

APRON ASSEMBLY

SECTION LA

APRON ASSEMBLY

Drg. Ref.	Part No.	Description	No. Off/Mc.	Drg. Ref.	Part No.	Description	No. Off/Mc.
LA5	A4729	Spring—Leadscrew Nut – – – –	1	LA38	A9782	Washer—Drive Shaft – – – –	1
LA6		Cap Hd. Screw—Leadscrew Nut (2 B.A. x 1¾")	1	LA39		Circlip—Drive Shaft (Anderton 1400—⅜") –	1
LA7	A2082	Gib Strip—Leadscrew Nut – – –	1	LA40	A9208	Knob Operating Spindle – – – –	1
LA8	A9193	Ch. Hd. Screw—Strip Securing – –	2	LA41	A9210	'Oilite' Bush – – – – –	2
LA9	A9194	Adjusting Screw—Gib Strip – –	1	LA42	A9211	'Oilite' Bush – – – – –	1
LA10	A9195	Adjusting Screw—Gib Strip – –	1	LA43	A9212/1	'Oilite' Bush—Flanged – – –	1
LA11	A9196	Leadscrew Guard – – – –	1	LA44	A7595	'Oilite' Bush – – – – –	1
LA12		Hex. Hd. Set Screw (2 B.A. x ½") – –	1	LA45	A9220	Clutch Insert – – – – –	1
LA13		Hex. Locknut (2 B.A.) – – – –	2	LA46	A9203/1	Stud—Gear Cluster – – – –	1
LA15	80002	Ball Knob (KB5/100) – – – –	1	LA47	65001	Oil Nipple (Tecalemit NC6057) – –	1
LA18	A9198	Hand Traverse Pinion – – –	1	LA48	10025/1	Apron Assembly (includes LA41, LA42, LA43)	1
LA19	65004	Sealing Plug—Apron (AQ330/15) – –	1	LA49		Cap Screw (M6 x 1 x 25 mm) – – –	4
LA20	70002	Woodruff Key (No. 404) – – –	1	LA50	10217	Thrust Washer – – – – –	1
LA21	A2087	Handwheel Assembly – – – –	1	LA51	10431	Circlip – – – – – –	1
LA22		Socket Set Screw (¼" B.S.F. x ¼") (Knurled		LA52	A9200/1	Bevel Pinion – – – – –	1
		Cup Point) – – – – –	1	LA53	A1975/3	Leadscrew Nut – – – – – set	1
LA23	A9199	Rack Pinion Assembly – – –	1	LA54	10508	Cam Peg – – – – – –	2
LA24	A2531	Oil Level Plug – – – – –	1	LA55	10528	Cam – – – – – –	1
LA25	65000	Oil Nipple (Tecalemit NC6055) – –	1	LA56	65007	'O' Ring (BS/USA115) – – –	1
LA27	A9201	Bevel Gear Cluster Assembly (includes LA44)	1	LA57	10529	Eccentric Sleeve – – – –	1
LA28	A9202	Thrust Washer – – – – –	1	LA58		Socket Set Screw (₅⁄₁₆" B.S.F. x ⅜",	
LA30		Socket Set Screw (¼" B.S.F. x ½") (Knurled				Half Dog Point) – – – –	1
		Cup Point) – – – – –	1	LA59	10530	Lever – – – – – –	1
LA31	A9204	Clutch Gear Assembly (includes LA45)	1	LA60		Socket Set Screw (2 B.A. x ¼", Cup Point) –	1
LA32	A9205	Drive Gear – – – – –	1	LA61	10424	Guard Plate (not illustrated) – – –	1
LA33	73010	Ball—Clutch (5 mm ∅) – – –	2				
LA34	A9206	Operating Spindle – – – –	1				
LA35		Circlip (Anderton 1400—⅜") – –	3				
LA36	A9207	Drive Shaft – – – – –	1				
LA37		C's'k Hd. Socket Screw (2 B.A. x ⅜") –	3				

Fig. 3.34 *Block diagrams: (a) drive train for a milling machine; (b) drive train for a centre lathe*

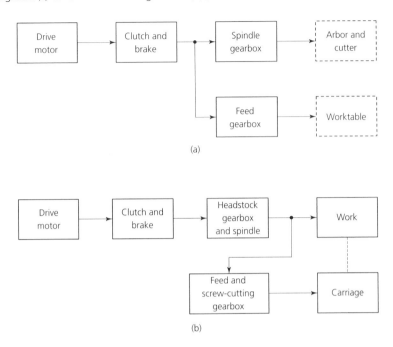

(a)

(b)

Fig. 3.35 *Installation drawing*

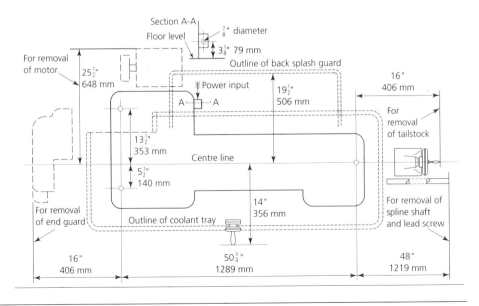

3.11 Pictorial views (oblique projection)

At the beginning of this chapter a pictorial view of a simple clamp was shown. That example was in isometric projection and we will be considering that method of drawing in the next section. For the moment, we will consider oblique projection since it is more simple.

Figure 3.36 shows a simple component drawn in oblique projection. The component is positioned so that you can draw one face true to size and shape. The lines running 'into' the page are called *receding* lines, which are usually drawn at 45° to the front face as shown. To improve the proportions of the drawing and make it look more realistic, the receding lines are drawn to half their true length. For ease of drawing the following rules should be observed:

- Any curve or irregular face should be drawn true shape (front view). For example, a circle on a receding face would have to be constructed as an ellipse, whereas if it were positioned on the front face it could be drawn with a compass.
- Wherever possible, the longest side should be shown on the front, true view. This prevents violation of perspective and gives a more realistic appearance.
- For long, circular objects such as shafts, the above two rules conflict. In this instance the circular section takes preference and should become the front view for ease of drawing, even though this results in the long axis receding.

Fig. 3.36 *Oblique drawing*

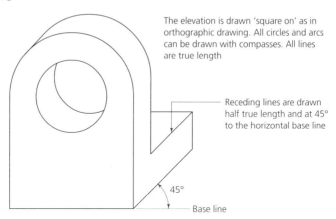

The elevation is drawn 'square on' as in orthographic drawing. All circles and arcs can be drawn with compasses. All lines are true length

Receding lines are drawn half true length and at 45° to the horizontal base line

45°

Base line

3.12 Pictorial views (isometric projection)

The bracket shown in Fig. 3.37 is drawn in isometric projection. The isometric axes are drawn at 30° to the base line. To be strictly accurate you should draw these receding lines to isometric scale and only the vertical lines are drawn to true scale. However, for all practical purposes, all the lines are drawn true length to save time.

Fig. 3.37 *Isometric drawing*

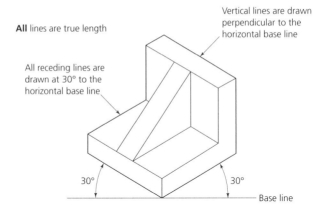

All lines are true length

Vertical lines are drawn perpendicular to the horizontal base line

All receding lines are drawn at 30° to the horizontal base line

30°

30°

Base line

Although an isometric drawing produces a more pleasing representation than an oblique drawing, it has the disadvantage that no curved profiles such as arcs, circles, radii, etc., can be drawn with a compass. All curved lines have to be constructed. You can do this by erecting a grid over the feature in orthographic projection, as shown in Fig. 3.38(a). You then draw a grid of equal size where it is to appear on the isometric drawing. The points where the circle cuts the grid in the orthographic drawing are transferred to the isometric grid, as shown in Fig. 3.38(b). You then draw a smooth curve through the points on the isometric grid and the circle appears as an ellipse. Figure 3.38(c) shows how this technique is applied to the example from the previous section. Note how the circle drawn with a compass in the oblique projection becomes an ellipse in the isometric projection. Despite the extra work, I prefer using isometric projection; it seems to look better. Figure 3.39 shows an example of a fully dimensioned isometric drawing.

Fig. 3.38 *Construction of isometric curves*

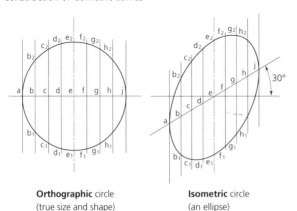

Orthographic circle
(true size and shape)

Isometric circle
(an ellipse)

1. Construct a grid over the true circle by dividing its centre line into an equal number of parts and erecting a perpendicular at each point.

2. Construct a similar grid on the isometric centre line.

3. Step off distances b_1-b-b_2, c_1-c-c_2, etc. on the isometric grid by transferring the corresponding distances from the true circle.

4. Draw a fair curve through the points plotted.

Fig. 3.39 *Fully dimensioned component in isometric projection. (Dimensions in millimetres)*

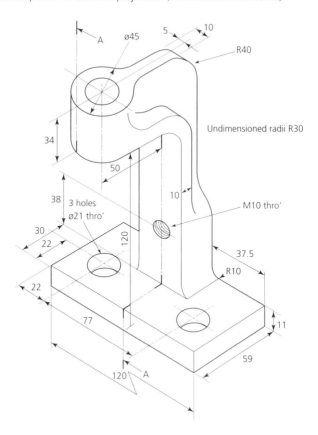

3.13 Sketching

We have now dealt with the types of drawing and drawing techniques you are most likely to meet with in an engineering workshop. The information given so far in this chapter should enable you, with practice, to interpret such drawings correctly. As has been stated previously, craftspersons are not expected to produce formal engineering drawings, they only have to know how to read them and interpret them correctly. However, they are expected to be able to produce sketches suitable for the manufacture of components – for example, the manufacture of replacement parts for maintenance purposes.

The same rules apply as for formal drawings, except that you are not going to use officially printed drawing paper, nor are you going to be sitting at a drawing board with a T-square or at a draughting machine. You will be sketching freehand on any convenient, clean piece of paper. However, it is handy to have some basic drawing aids available, possibly saved from your school geometry set. Let us see what you need.

3.13.1 *Writing equipment*

Pencils

Ordinary wooden pencils have a habit of becoming blunt or breaking at the most inconvenient moments, particularly if they are soft enough to leave a clear outline. It is better to use *clutch type* pencils that are specially made for drawing. The lead has a constant, precise diameter and it is supported right to its point. Various diameters of lead are available and suitable sizes are 0.3 mm for thin lines such as construction lines and dimension lines, and 0.5 mm or 0.7 mm for outlines. Pencil lines have the advantage that they can be rubbed out if you make a mistake. Unfortunately they are not as permanent as ink lines. You will also require a suitable plastic eraser.

Pens

Drawing pens with disposable ink cartridges are widely used by professional draughtspersons if they are still drawing manually. These have the advantage that the nib unit can be detached from the pen body and fitted into a compass so that arcs and circles can be drawn without any change in line texture. However, such pens are more suitable for regular use and require careful cleaning and maintenance if they are to be kept in good condition. For occasional sketching, disposable *graphic liner* type pens are more suitable. They do not have to be cleaned after use and, providing you remember to replace the cap, they will not dry up. They produce a constant width, dense black line that prints well and will not fade. Like clutch pencils they are available in a variety of line widths, and the sizes recommended for the pencils are equally appropriate for these pens.

3.13.2 *Drawing instruments*

Although the art of sketching assumes the drawing will be done freehand, until you are experienced it helps to have some instruments near at hand. These basic instruments are shown in Fig. 3.40:

- A rule is essential for measuring and drawing straight lines. You also need to be able to measure in order to dimension your drawings and keep them to scale.

Fig. 3.40 *Drawing instruments*

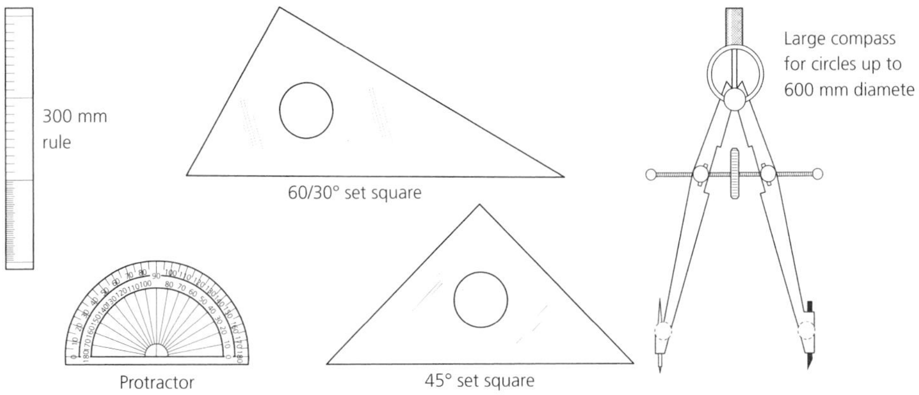

300 mm rule

60/30° set square

Large compass for circles up to 600 mm diameter

Protractor

45° set square

- A 45° set square is used for drawing chamfers to remove sharp corners from components, for section shading, and for drawing the receding lines if you are drawing in oblique projection.
- A 60/30° set square is used for isometric drawing and also to provide an alternative angle for section shading if the outline of the component contains a 45° angle.
- A protractor for all those angles not provided by the set squares.
- A compass for drawing circles.

3.13.3 *Orthographic sketching*

Figure 3.41 shows a bracket which has to be made and fitted to the end of a machine tool bed. Figure 3.42 shows how you make a freehand orthographic sketch for the bracket. In fact it is a freehand detail drawing.

- Use a sheet of clean, flat, good-quality paper of adequate size and an 'H' grade pencil. You also want a clean, flat surface to draw on. A clip board is handy.
- Paper feintly ruled with squares is helpful as it gives you a guide for lines at right angles to each other.
- Now make outline sketches of the views you require using thin, feint lines.
- When you are satisfied with your initial sketches, draw in the outline more heavily and add any necessary details.
- When the basic sketch is complete check it for the omission of any essential details.
- Having completed the outline, you now have to add the dimensions. Use a rule or other instruments such as micrometer and vernier calipers to take the measurements you require and transfer them to the drawing. If the measurements are taken accurately and shown correctly on your sketch, your sketch does not have to be to scale. In any case a drawing should never be scaled since you cannot draw to the accuracy required for the manufacture of an engineering component.
- Make enlarged sketches of any small details that cannot be clearly shown in the main views.
- Although it is only a sketch for use once, it must incorporate all the rules, information and conventions discussed earlier in this chapter. Only then will the person making the component be able to make it correctly.

Fig. 3.41 *Bracket*

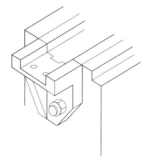

Fig. 3.42 *Orthographic sketching*

FRONT
SIDE
PLAN

Use cleanest paper available and a sharp pencil or ballpoint pen. Rest the work on a flat surface or support it firmly. The use of paper ruled with squares is helpful.

Attempt to keep the drawing as clean and clear as possible.

Make rough shetches to decide what views are necessary.

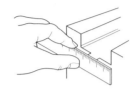

Take measurements and mark on sketch.

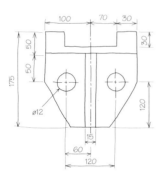

If measurements are taken accurately, and shown correctly, a freehand sketch need not be exactly in proportion. It must be clear and neat to prevent error in reading from it. Symbols used should adhere to BS 308.

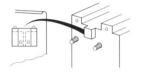

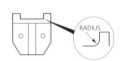

When shapes are complete, check carefully against existing objects for omissions.

Make an enlarged sketch of detail where necessary.

3.13.4 *Pictorial sketching*

Pictorial sketches can be in oblique or in isometric projection. Figure 3.43 shows you how to make a sketch in *oblique* projection.

- Step 1: sketch 'boxes' to contain the outline of the finished drawing.
- Step 2: lightly sketch in the details of the component or assembly being drawn.
- Step 3: go over the outline more heavily to make it stand out.
- Step 4: remove any construction lines that may cause confusion.
- Step 5: add dimensions as required.

Figure 3.44 shows you how to make a sketch of a two-jaw chuck in *isometric* projection. The technique is similar to that used for the previous, oblique sketch. However, the initial outline 'boxes' are drawn in isometric projection.

- Step 1: sketch the outline boxes in isometric projection as shown.
- Step 2: now sketch the curves using feint lines. Remember that in isometric projection these will be ellipses or parts of ellipses. A template may be helpful.

Fig. 3.43 *Pictorial sketching (oblique): Step 1 – general outlines (very feint); Steps 2 and 3 – add details and line in*

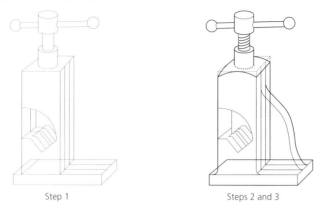

Step 1   Steps 2 and 3

Fig. 3.44 *Pictorial sketching (isometric): Step 1 – general outlines (very feint); Step 2 – add details and line in*

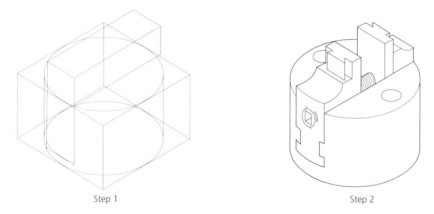

Step 1   Step 2

- Step 3: add any detail that is required.
- Step 4: finally line in the outline more heavily and remove any construction lines that may cause confusion.
- Step 5: add dimensions as required.

3.1 *First- and third-angle projection*
  (a) State which of the examples shown in Fig. 3.45 are in **first-** or **third**-angle projection.
  (b) Copy and complete the examples shown in Fig. 3.46. The projection symbol is placed below each example for your guidance. (*Note*: sometimes complete views are missing, sometimes only lines and features.)

Fig. 3.45 *Exercise 3.1(a)*

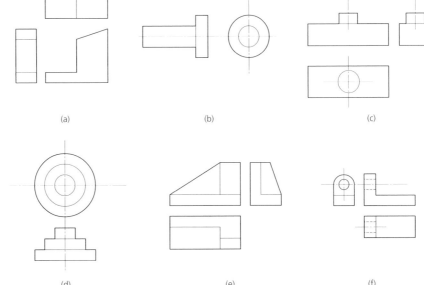

(a) (b) (c)

(d) (e) (f)

Fig. 3.46 *Exercise 3.1(b)*

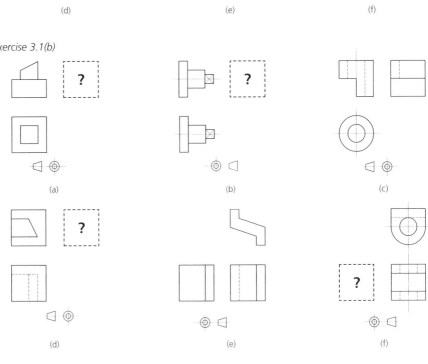

(a) (b) (c)

(d) (e) (f)

3.2 *Types of line*

(a) Copy and complete Fig. 3.47.

(b) With the aid of a sketch show what is meant by the terms:

(i) dimension line (ii) leader line (iii) projection line

Fig. 3.47 *Exercise 3.2(a)*

Line	Description	Application
————	Continuous bold	
————	Continuous fine	
∼∼∼∼		Limit of partial view
- - - - - - - -	Fine short dashes	
— · — · —		Centre lines
	Fine chain, bold at ends and changes of direction	Cutting planes

(c) With reference to exercise (b):
  (i) state the type of line that should be used
  (ii) indicate on your sketch where a short extension is required and where a small gap is required

3.3 *Dimensioning*
(a) With the aid of sketches show how a simple component can be dimensioned from:
  (i) a pair of mutually perpendicular datum edges (or surfaces)
  (ii) a datum line
  (iii) a datum point
(b) With the aid of sketches show how you should dimension the following features:
  (i) a circle (show **four** methods of dimensioning and use the diameter symbol)
  (ii) a radius (both convex and concave)
  (iii) an angle or chamfer
(c) Figure 3.48 shows a number of abbreviations found on engineering drawings. Copy and complete the figure.

Fig. 3.48 *Exercise 3.3(c)*

Name	Abbreviation	Sketch
	S'FACE	
	C'BORE	
	C'SK	

3.4 *Conventions*
(a) Give **two** reasons why standard conventions are used on engineering drawings.
(b) Figure 3.49 relates to some commonly used drawing conventions. Copy and complete the figure.

Fig. 3.49 *Exercise 3.4(b)*

Title	Subject	Convention
External screw threads (detail)	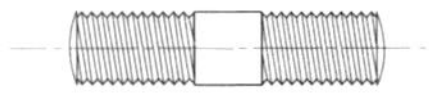	
Internal screw threads (detail)	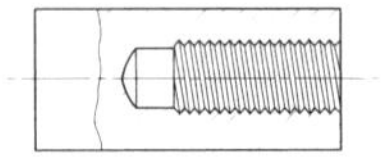	
Diamond knurling	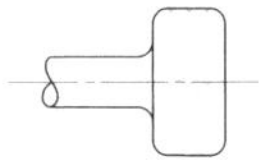	
Square on shaft	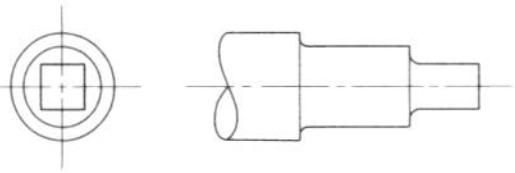	
Holes on circular pitch	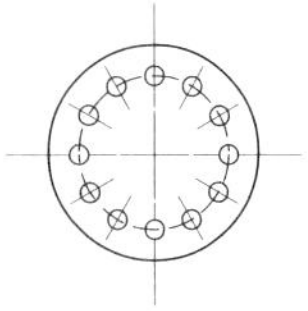	
Ball and roller bearings	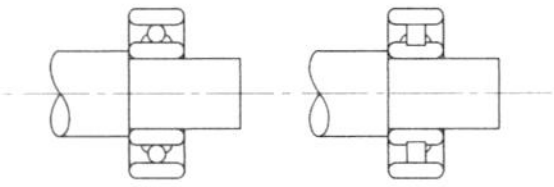	
Splined shafts	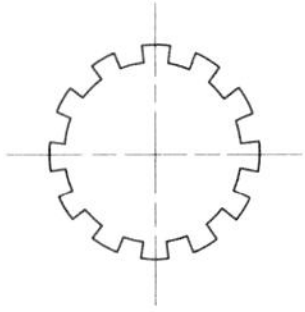	
Cylindrical tension spring	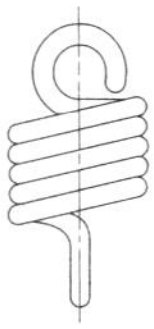	
Repeated parts	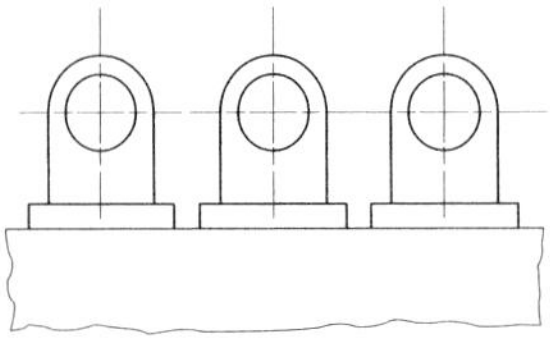	

3.5 *Sectioning*

    (a) Figure 3.50 shows a drawing that has been incorrectly sectioned. Redraw it **correctly** sectioned.

    (b) With reference to Fig. 3.50, draw a section on the cutting plane *AA*.

Fig. 3.50 *Exercise 3.5*

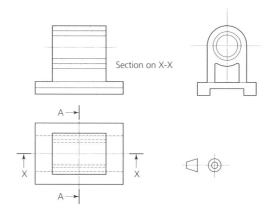

3.6 *Pictorial views*

    (a) Figure 3.51 shows a simple workpiece. Sketch it in:

        (i) oblique projection

        (ii) isometric projection

Fig. 3.51 *Exercise 3.6(a). (Dimensions in millimetres)*

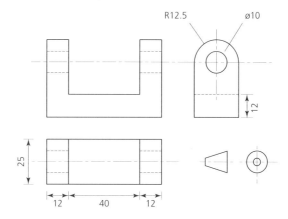

    (b) Figure 3.52 shows a component in isometric projection. Sketch it full-size in:

        (i) *first-angle* orthographic projection with the necessary dimensions so that your sketch can be used as a detail drawing for the manufacture of the component

(ii) *third-angle* orthographic projection with the necessary dimensions so that your sketch can be used as a detail drawing for the manufacture of the component

Fig. 3.52 *Exercise 3.6(b). (Dimensions in millimetres)*

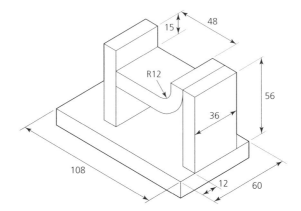

4 Measuring

When you have read this chapter, you should understand:

- What is meant by dimensional properties.
- What is meant by linear measurement.
- How to make linear measurements.
- What is meant by angular measurement.
- How to make angular measurements.
- The factors affecting accuracy of measurement.
- The terminology of measurement.
- The general rules for accurate measurement.
- How to measure workpieces having square, rectangular, circular and irregular-shaped sections.

4.1 Introduction

Measuring can be considered to be the most important process in engineering. Without the ability to measure accurately, we cannot:

- Mark out components as described in Chapter 5.
- Set up machines correctly to produce components to the required size and shape.
- Check components whilst we are making them to ensure that they finally end up the correct size and shape.
- Inspect finished components to make sure that they have been manufactured correctly.

In this chapter we will first decide what needs to be measured (dimensional properties) and then we will look at the different ways in which we can measure these dimensional properties, the equipment required, and the care of that equipment.

4.2 Dimensional properties

When checking or measuring a component there are a number of dimensional properties which you have to consider. Let us now consider the more important ones.

4.2.1 *Length*

When you measure length, you measure the shortest distance in a straight line between two points, lines or faces. It doesn't matter what you call this distance (width, thickness, height, breadth, depth, height, diameter), it is still a measurement of length. There are two systems of measurement for length. These are based upon the standard yard and on the standard metre. These standards will be considered in Section 4.3.

4.2.2 *Flatness*

A flat surface lies in a true plane. However, be careful; appearances can be deceptive. Figure 4.1 shows a surface that is definitely not flat. Yet, if you checked along the straight lines parallel to the edges with a straight edge, it would appear to be flat. We call these straight lines the *generators*. For the surface to be flat, these generators must:

- Be straight.
- Lie in the same plane.

In our example, they are straight but they do not lie in the same plane. This would be shown up by placing a straight edge across the surface from corner to corner, diagonally.

Fig. 4.1 *Lack of flatness generated by straight lines and difficult to detect with a straight edge*

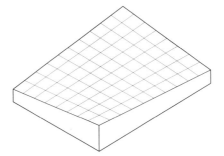

When you check the flatness of a component, you are measuring by how much the surface of the component deviates from a true plane. There are various ways of measuring flatness, including the use of sophisticated optical and electronic instruments. In the workshop it is usually adequate to compare the surface of the component with a surface plate. These are available in various grades of accuracy. Figure 4.2 shows how you can compare a component with a surface plate and how you can remove the local high spots with a scraper. Strictly, this is not a measuring process since no measurements are taken. However, it is a means of arriving at a reasonably flat surface. This is the procedure:

- A small amount of engineer's blue is smeared lightly onto the surface plate.
- The component is slid gently over the surface plate (Fig. 4.2(a)).

- Any high spots on the component will pick up some of the blue (Fig. 4.2(b)).
- These high spots are removed with a scraper (Fig. 4.2(c)).
- The process is repeated until the component is uniformly covered with a fine pattern of high spots (Fig. 4.2(d)).

Fig. 4.2 *Producing a flat surface: (a) workpiece being compared with a surface plate; (b) initial appearance of surface to be scraped; (c) using a scraper to remove the high spots; (d) appearance of surface after scraping*

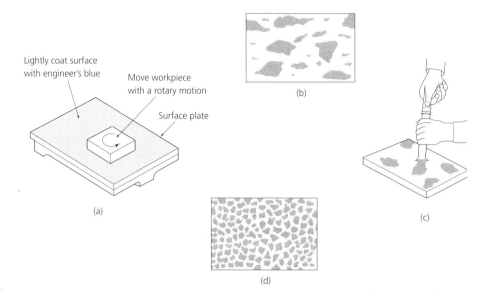

This technique has been in use since the earliest days of engineering. It was in use long before the introduction of optical and electronic measuring equipment. So where did the first surface plate originate? In fact engineers started with three surface plates. This is because flat (plane) surfaces can be produced by working three surface plates together, as shown in Fig. 4.3. The three plates were made as flat as possible by whatever means was available. Plate A was then compared with plate B and they were scraped until their

Fig. 4.3 *Comparison of three surface plates to achieve flatness*

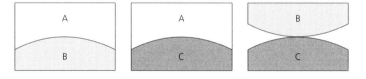

surfaces appeared to be flat when compared with each other. Plate A was then compared with plate C. This showed up apparent errors in plate C, which was then scraped 'flat' when compared with plate A. However, when plates B and C were compared any deviations from a plane surface were immediately apparent. The plates were repeatedly interchanged and scraped until they were flat when compared in any combination. That is, the surfaces of A and B, A and C, and B and C only showed overall contact when perfect flatness had been achieved.

Narrow surfaces (edges) can be checked with a straight edge.

- A plain straight edge (Fig. 4.4(a)) can be considered to be a long narrow surface plate and used with engineer's blue as described above
- A bevelled straight edge can be used by checking the light gap, as shown in Fig. 4.4(b).

Fig. 4.4 *Straight edges: (a) plain straight edge; (b) bevelled straight edge*

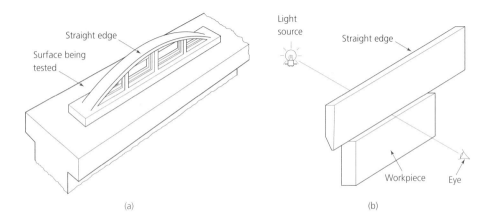

4.2.3 *Parallelism*

This is the constancy of distance between two lines or surfaces. Parallelism can be measured in various ways depending upon the job in hand. Figure 4.5(a) shows how an indicating instrument (micrometer) can be used. If it shows the same reading at both ends of the work, then the work is parallel. Figure 4.5(b) shows how a non-indicating instrument (caliper) is used. The work is parallel if the 'feel' is the same at both ends. Parallelism also applies to rectangular components. Figure 4.5(c) shows how a micrometer and a depth micrometer can be used to check for parallelism. In each example the reading or the 'feel' must be the same at both ends of the component. Figure 4.5(d) shows two ways in which a surface gauge (scribing block) can be used to check for parallelism.

Fig. 4.5 *Checking for parallelism: (a) with an indicating instrument (micrometer); (b) with a non-indicating instrument (caliper); (c) rectangular components; (d) with a scribing block (surface gauge)*

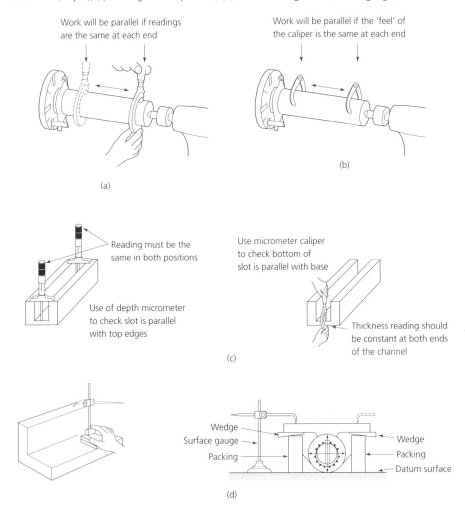

4.2.4 *Surface texture*

This is the surface condition of a component after casting, forming or cutting. The difference between roughness and waviness is shown in Fig. 4.6(a), and the terminology of surface texture is shown in Fig. 4.6(b). Surface texture (also called surface roughness) can be assessed by using comparison blocks or a surface-measuring machine. Surface comparison blocks come as a boxed set and represent typical surface roughness values as produced by various workshop processes. Comparison is made by drawing your finger-nail across the component and then across the test blocks until a match of feel is achieved. This requires a lot of skill and is not very accurate.

The surface-measuring machine draws a measuring head across the surface being checked. The stylus of the measuring head (which is similar to a record player pickup head) is deflected by the surface roughness. These deflections generate minute electrical signals that are amplified electronically and printed out on a paper tape. The tape is pre-printed with a continuous grid. The trace is an enlarged reproduction of the component surface condition. Knowing the amplification factor of the equipment, the actual value of the peaks and troughs can be scaled and measured. Figure 4.7(a) shows a typical surface-texture-measuring machine and Fig. 4.7(b) shows a typical printout from such a machine.

Fig. 4.6 *Surface texture: (a) roughness and waviness; (b) surface texture terminology*

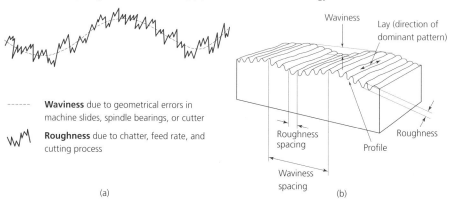

(a)

(b)

Fig. 4.7 *Surface texture measurement: (a) electronic surface texture assessment equipment; (b) typical surface texture traces (source: Taylor Hobson Ltd)*

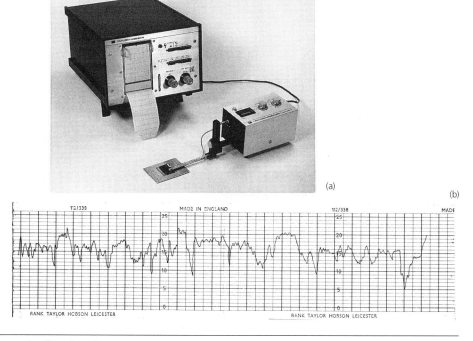

(a)

(b)

4.2.5 *Angles*

An angle is a measure of the inclination of one line or one surface to another. Right angles may be checked by the use of a try-square and angles other than a right angle may be measured by a protractor. These instruments are introduced in Section 4.3 of this chapter. Alternatively, tapers (cones) may be checked by the use of taper plug and taper ring gauges, and examples of these are shown in Section 4.4.

4.2.6 *Profiles*

The profile of a component is its outline shape. You can check simple shapes such as a rectangle by using a try-square, or a radiused corner by use of radius gauges, as shown in Fig. 4.8(a). More complex shapes can be checked by use of a template, as shown in Fig. 4.8(b). Where greater accuracy is required and where the profile is more complex, an enlarged image of the component profile can be projected onto the screen of an optical projector. The profile projected onto the screen is compared with enlarged templates or transparent overlays, as shown in Fig. 4.8(c). Electronic three-dimensional profile-measuring machines are also available.

Fig. 4.8 *Checking a profile: (a) checking a radius with a radius gauge; (b) use of template to turn a profile; (c) use of a profile projector to check a screw thread form*

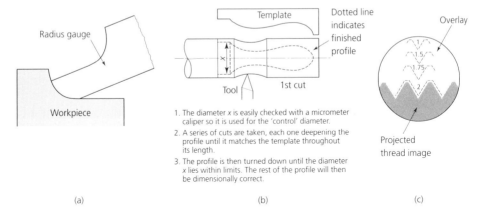

(a)   (b)   (c)

4.2.7 *Relative position*

The alignment of major components and subassemblies prior to final assembly and fixing can be carried out in various ways. Horizontal and vertical planes can be established by the use of precision bubble levels (spirit levels), as shown in Fig. 4.9(a). For this purpose, the small levels built into the various measuring heads of a combination set are not sufficiently accurate.

Alignment in the vertical plane can also be achieved using a plumb-bob, as shown in Fig. 4.9(b). A plumb-bob always hangs vertically: that is, it always points towards the centre of the Earth. Since plumb-bobs are radial to the Earth's centre, no two plumb-bobs can be truly parallel to each other. However, for all practical purposes, the radius of the Earth is so large that this lack of parallelism can be ignored over short distances.

Optical alignment can also be used. Figure 4.9(c) shows how simple arrangement of a light source and sights may be used. Where greater accuracy is required, such sophisticated instruments as alignment telescopes, autocollimators and lasers may be used and, although you may see these in use at your place of work, they are outside the scope of this book.

Fig. 4.9 *Checking for alignment: (a) use of precision level to test for mutual perpendicularity (squareness) of slideways; (b) use of plumb-bob to test for axial alignment; (c) optical alignment*

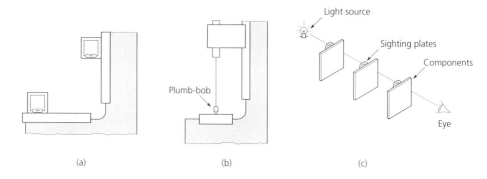

4.2.8 *Roundness*

A round component is one with a truly circular cross-section; that is, a section through a perfect cylinder or a perfect cone. Roundness implies how closely a cylindrical or conical component approaches this ideal. You can measure the diameter of a circular component using a micrometer or a vernier caliper. Such measurements cannot detect out-of-roundness, such as *ovality*. This is shown in Fig. 4.10(a).

Some forms of out-of-roundness cannot be measured in this way. Figure 4.10(b) shows the problems that arise when measuring a component which has constant diameter, lobed cross-section. The only way to detect this form of out-of-roundness is by use of a dial test indicator (DTI) mounted on a comparator stand with the work being rotated beneath it in a V-block, as shown in Fig. 4.10(c).

4.2.9 *Concentricity*

Concentricity implies a number of diameters having a common axis. For example, if you draw a number of circles of different diameters on a piece of paper without removing the point of your compass, then all the circles will be *concentric*.

Fig. 4.10 *Testing for out-of-roundness: (a) ovality is indicated if measurement d₁ and d₂, taken at right angles, differ; (b) some types of lobed cross-section will not show up when checked on a flat anvil; (c) if the component is rotated in a vee-block, any out-of-roundness will immediately show up as the component rides up and down*

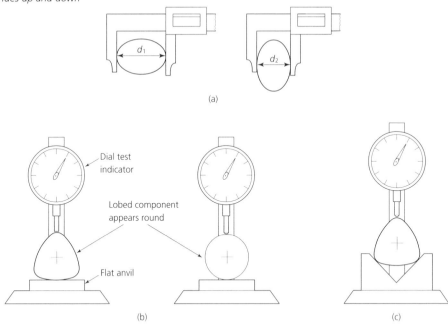

Figure 4.11(a) shows that a bush with concentric diameters will have a constant wall thickness. However, if it does not have a constant wall thickness its diameters will not be concentric, they will be *eccentric*. To test this bush for concentricity, it is rotated on a cylindrical mandrel under a DTI. If the wall thickness is constant the reading of the DTI will also be constant. The bush does not have to be a good fit on the mandrel, there can be plenty of clearance between the bore of the bush and the mandrel. It will not affect the accuracy of the test.

Fig. 4.11 *Testing for concentricity: (a) constant wall thickness; (b) solid component; (c) bored component*

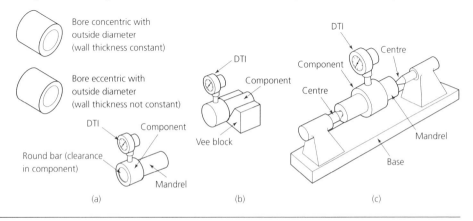

Figure 4.11(b) shows how a solid object can be tested for concentricity using a V-block and DTI. Again concentricity will be indicated by no movement in the reading of the DTI. Figure 4.11(c) shows a bush that is a light push fit on a mandrel. The mandrel is mounted between bench centres and rotated under a DTI. Any eccentricity is shown up by variations in the DTI reading.

4.2.10 *Accuracy of form*

This is the combination of all the dimensional properties we have considered so far. For many components it is only necessary to specify toleranced dimensions. The machining process can then be relied upon to provide the required accuracy of form. However, the precision and performance criteria for engineering components is becoming increasingly exacting and accuracy of form cannot always be assumed. For this reason, geometrical as well as dimensional tolerances are being increasingly used on the drawings for precision components. Although beyond the scope of this book, if you are interested, you will find the symbols for geometrical tolerancing and examples of their use in BS 308: Part 3. An example of geometrical tolerancing is shown in Fig. 4.12. Now that we have decided what has to be measured, let's look at how the measurement processes are performed.

Fig. 4.12 *Geometrical tolerancing*

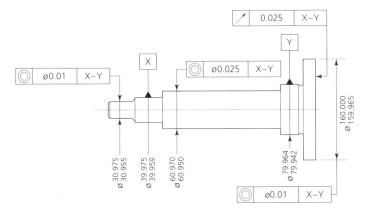

4.3 Linear measurement

4.3.1 *Background*

Take a rule and measure this page from top to bottom and then from side to side. I could then ask you what you have just done. Your answers would be correct if you replied that you had just measured the size of the page, or that you had just used a rule. What I would have preferred, is for you to have replied that you had just *compared* the size of the page with the scale on the rule. All measurements are *comparative processes*. Your rule was engraved on a machine. The engraving machine, in turn, was tested by comparing it with some more accurate standard. This, in turn, would be tested against an even more accurate standard, and so on until we get to the ultimate standard from which all other measuring instruments are derived.

Over the years many standards of length have been used. In the UK we used the *imperial standard yard*. In America they used the America standard yard, which was slightly larger. After 1945 these were harmonised to become the *international standard yard*, part way between the two. Other countries used variations of the French *standard metre*.

In order that world trade could flourish it became necessary for national standards to be 'harmonised'. In the UK this is the responsibility of the British Standards Institution (BSI) which works in conjunction with the International Standards Organisation (ISO) and European Community standards committees. Such international standardisation is essential to ensure the interchangeability of components and equipment manufactured in different countries. Standardisation refers to both dimensional accuracy and quality (fitness for purpose). The units of the *Système Internationale d'Unités* (SI) are now used throughout the world.

The fundamental unit of length is the *metre* and, currently, this is defined as:

The length of a path travelled by laser light in 1/299 792 458 seconds. The light being realised through the use of an iodine stabilised helium-neon laser.

The international standard yard is defined as 0.9144 metre. Whilst units based on the metre are used world wide, units based on the international standard yard are mainly used in the UK and the USA. The relationships between most of the units of linear measurement you are likely to meet in engineering are shown below. Remember that although the metric system is becoming predominant in the UK, the imperial (inch) system is still widely used. Both systems will be shown in this chapter.

12 inches	=	1 foot
3 feet	=	1 yard
1 yard	=	0.9144 metres
1000 millimetres	=	1 metre
25.4 millimetres	=	1 inch

Note: 'Yard' = 'International standard yard'.

4.3.2 *Steel rules*

Use of steel rules

The steel rule is frequently used in workshops for measuring components of limited accuracy quickly. The quickness and ease with which it can be used, coupled with its low cost, makes it a popular and widely used measuring device. Metric rules may be obtained in various lengths from 150 to 1000 mm (1 metre). Imperial rules may be obtained in various lengths from 6 to 36 inch (1 yard). Figure 4.13 shows examples of rule scales in both systems. It is convenient to use a rule engraved in both systems, one system on the front and the other on the back.

Fig. 4.13 *Steel rule scales – the reverse side shows: 1/64 inch; 1/10, 1/20, 1/100 inch*

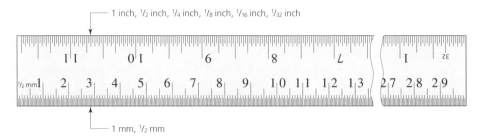

1 inch, ½ inch, ¼ inch, ⅛ inch, ¹⁄₁₆ inch, ¹⁄₃₂ inch

1 mm, ½ mm

Steel rules may be 'rigid' or 'flexible' depending upon their thickness and the 'temper' of the steel used in their manufacture. When choosing a steel rule the following points should be looked for. It should be:

- Made from hardened and tempered, corrosion-resistant spring steel.
- Engine divided – that is, the graduations should be precision engraved into the surface of the metal.
- Ground on the edges so that it can be used as a straight edge when scribing lines or testing a surface for flatness.
- Ground on one end so that this end can be used as the zero datum when taking measurements from a shoulder.
- Satin chrome finished to reduce glare and make the rule easier to read, and to prevent corrosion.

No matter how accurately a rule is made, all measurements made with a rule are of limited precision. This is because of the difficulty of sighting the graduations in line with the feature being measured. Some ways of minimising sighting errors are shown Fig. 4.14.

Fig. 4.14 *Use of a rule – measuring the distance between: (a) two scribed lines; (b) two faces using a hook rule; (c) two faces using a steel rule and a tool bit as an abutment*

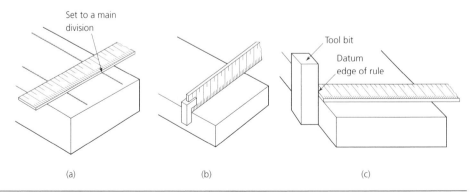

Set to a main division

Tool bit

Datum edge of rule

(a)   (b)   (c)

When using a rule to make direct measurements, as in Fig. 4.14, the accuracy of measurement depends upon the *visual alignment* of a mark or surface on the work with the corresponding graduation on the rule. This may appear relatively simple but, in practice, errors can very easily occur, as shown in Fig. 4.15. These errors can be minimised by using a thin rule and keeping your eyes directly above and at 90° to the mark on the work. If you look at the work and the rule from an angle, you will get a false reading. This is known as *parallax* error. Reference to Fig. 4.15 shows that:

- If you are looking at the mark 'M' on the work from position 'L' you will read the rule graduation at 'l'.
- If you are looking at the mark 'M' on the work from position 'R' you will read the rule graduation at 'r'.
- Only if you are looking at the mark on the work from position 'A' immediately above the mark and at 90° to it, will you get the true reading at 'a' on the rule scale.

Fig. 4.15 *Effect of parallax*

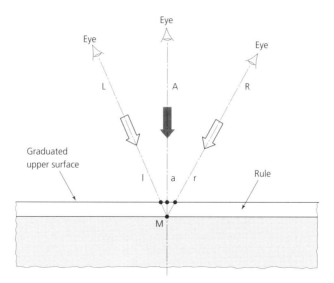

Care of steel rules
A good rule should be looked after carefully to prevent wear and damage to its edges and to the datum end. It should never be used as a scraper or a screwdriver, and it should never be used to clean the swarf out of the T-slots in the tables of machine tools. After use, plain steel rules should be lightly oiled to prevent rusting. Dulling of the surfaces will make the scales difficult to read. There is no need to oil satin-chrome-plated rules, just wipe them clean.

4.3.3 *Line and end measurement*

Linear distances sometimes have to be measured between two lines, sometimes between two surfaces and sometimes between a combination of line and surface. Measurement between two lines is called *line measurement*. Measurement between two surfaces is called *end measurement*. It is difficult to convert between end systems of measurement and line systems of measurement and vice versa. For example, a rule (which is a line-system-measuring device) is not convenient for the direct measurement of distances between two edges, Similarly, a micrometer (which is an end system measuring device) would be equally inconvenient if used to measure the distance between two lines. The measuring device must always be chosen to suit the job in hand.

4.3.4 *Calipers and their use*

Calipers are used in conjunction with a rule to transfer the distance across or between the faces of a component in such a way as to reduce sighting errors – that is, to convert from *end measurement* to *line measurement*. Firm-joint calipers are usually used in the larger sizes and spring-joint calipers are used for fine work. Examples of internal and external calipers of both types are shown in Fig. 4.16 together with examples of their uses.

The accurate use of calipers depends upon practice, experience and a highly developed sense of feel. When using calipers, the following rules should be observed:

- Hold the caliper gently and near the joint.
- Hold the caliper square (at right angles) to the work.

Fig. 4.16 *Construction and use of calipers*

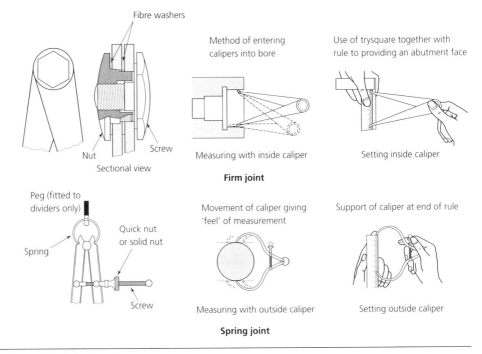

- No force should be used to 'spring' the caliper over the work. Contact should only just be felt.
- The caliper should be handled and laid down gently to avoid disturbing the setting.
- Lathe work should be *stationary* when taking measurements. This is essential for *safety* and *accuracy*.

4.3.5 *Micrometer caliper*

Use of the micrometer caliper

Most engineering work has to be measured to much greater accuracy than it is possible to achieve with a rule, even when aided by the use of calipers. To achieve this greater precision, we have to use measuring equipment of greater accuracy and sensitivity. One of the most familiar measuring instruments used in engineering workshops is the *micrometer caliper*. This is frequently referred to as a 'micrometer' or simply a 'mike'. The constructional details of a typical micrometer caliper is shown in Fig. 4.17.

The operation of this instrument depends upon the principle that the distance a nut moves along a screw is proportional to the number of revolutions made by the nut and the lead of the screw thread. Therefore by controlling the number of complete revolutions made by the nut and the fractions of a revolution made by a nut, the distance it moves along the screw can be accurately controlled. It does not matter whether the nut rotates on the screw or the screw rotates in the nut, the principle of operation still holds good.

In a micrometer caliper, the screw thread is rotated by a thimble which has a scale that indicates partial revolutions. The barrel of the instrument has a scale which indicates

Fig. 4.17 *The micrometer caliper (source: Moore and Wright Ltd)*

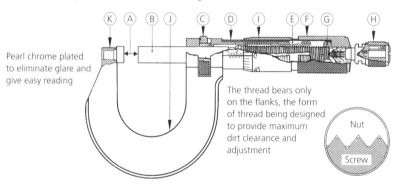

Pearl chrome plated to eliminate glare and give easy reading

The thread bears only on the flanks, the form of thread being designed to provide maximum dirt clearance and adjustment

A *Spindle and anvil faces* – Glass hard and optically flat, also available with tungsten carbide faces
B *Spindle* – Thread ground and made from alloy steel, hardened throughout, and stabilised
C *Locknut* – Effective at any position. Spindle retained in perfect alignment
D *Barrel* – Adjustable for zero setting. Accurately divided and clearly marked, pearl chrome plated
E *Main nut* – Length of thread ensures long working life
F *Screw adjusting nut* – For effective adjustment of main nut
G *Thimble adjusting nut* – Controls position of thimble
H *Ratchet* – Ensures a constant measuring pressure
I *Thimble* – Accurately divided and every graduation clearly numbered
J *Steel frame* – Drop forged
K *Anvil end* – Cutaway frame facilitates usage in narrow slots

'whole' revolutions. In a standard metric micrometer caliper the screw has a lead of 0.5 mm and the thimble and barrel are graduated, as in Fig. 4.18.

Fig. 4.18 *Micrometer scales (metric)*

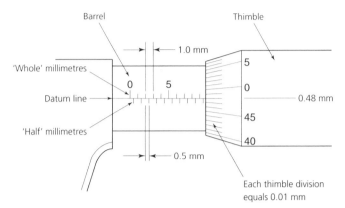

Since the lead of the screw of a standard metric micrometer is 0.5 mm and the barrel divisions are 0.5 mm apart, one revolution of the thimble moves the thimble along the barrel a distance of one barrel division (0.5 mm). The barrel divisions are placed on alternate sides of the datum line for clarity. Further, since the thimble has 50 divisions and one revolution of the thimble equals 0.5 mm, then a movement of *one thimble division* equals 0.5 mm/50 divisions = 0.01 mm.

A metric micrometer caliper reading is given by:

- The largest visible 'whole' millimetre graduation visible on the barrel, *plus*
- The next 'half' millimetre graduation, if this is visible, *plus*
- The thimble division coincident with the datum line.

Therefore the micrometer scales shown in Fig. 4.18 read as follows:

9 'whole' millimetres	=	9.00
1 'half' millimetre	=	0.50
48 hundredths of a millimetre	=	0.48
	=	9.98 mm

Figure 4.19 shows the scales for a micrometer graduated in 'inch' units. The micrometer screw has 40 T.P.I. (threads per inch), therefore the lead of the screw is 1/40 inch (0.025 inch). The barrel graduations are 1/10 inch subdivided into 4. Therefore each subdivision is 1/40 inch (0.025 inch) and represents one revolution of the thimble. The thimble carries 25 graduations. Therefore one thimble graduation equals a movement of 0.025 inch/25 = 0.001 inch. This is one thousandth part of an inch and is often referred to by engineers as a 'thou'. Thus 0.015 inch could be referred to as 15 'thou'.

An inch micrometer reading is given by:

- The largest visible 1/10 inch (0.1 inch) division, *plus*
- The largest visible 1/40 inch (0.025 inch) division, *plus*
- The thimble division coincident with the datum line.

Fig. 4.19 *Micrometer scales (English)*

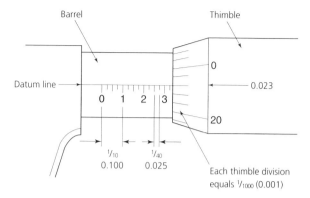

Therefore, the micrometer scales shown in Fig. 4.19 read as follows:

$$
\begin{array}{ll}
\text{3 tenths of an inch} & = 0.300 \\
\text{1 fortieth of an inch} & = 0.025 \\
\text{23 thousandths of an inch} & = \underline{0.023} \\
& = 0.348
\end{array}
$$

Figure 4.20 shows some further examples of English (inch) and metric micrometer scales and their readings. Try to work them out for yourself before looking at the correct readings given at the end of this chapter (page 139).

Fig. 4.20 *Micrometer caliper reading exercises*

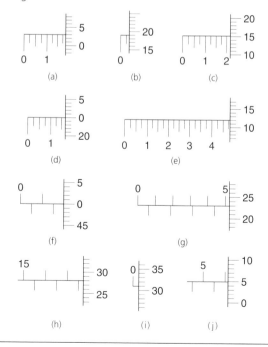

Care of the micrometer caliper

Unless a micrometer caliper is properly looked after it will soon lose its initial accuracy. To maintain this accuracy you should observe the following precautions:

- Wipe the work and the anvils of the micrometer clean before making a measurement.
- Do not use excessive measuring pressure, two 'clicks' of the ratchet are sufficient.
- Do not leave the anvil faces in contact when not in use.
- When machining, stop the machine before making a measurement. Attempting to make a measurement with the machine working can ruin the instrument and also lead to a serious accident. This rule applies to all measuring instruments and all machines.

From time to time you should check a micrometer caliper and make any necessary adjustment using the special spanner found in its case. The procedures for making the adjustments are as follows:

- Any looseness in the screw can be taken up by a slight turn of the screw-adjusting nut. (Refer to item 6 in Fig. 4.17.)
- Zero error. Periodically the anvil faces should be wiped clean and carefully closed with normal measuring pressure (two 'clicks' of the ratchet). If the zero graduation of the thimble does not coincide with the datum line on the barrel, turn the barrel in the frame using the 'C' end of the spanner until the datum line comes into line with the zero graduation of the thimble. (Refer to item 4 in Fig. 4.17.)

Although easy to read and convenient to use, micrometer calipers have two disadvantages:

- A limited range of only 25 mm. Thus a range of micrometers are required, for example:

    0–25 mm, 25–50 mm, 50–75 mm, and so on.

- Separate micrometers are required for internal and external measurements. The micrometer caliper so far described can only be used for external measurements.

4.3.6 *Internal micrometer*

An internal micrometer is shown in Fig. 4.21(a). It is used for measuring bore diameters and slot widths from 50 mm to 210 mm. For any one extension rod its measuring range is 20 mm. A range of extension rods in stepped lengths are provided in the case with the measuring head. It suffers from two important limitations.

- It cannot be used to measure small holes less than 50 mm diameter.
- It cannot be easily adjusted once it is in the hole and this affects the accuracy of contact 'feel' that can be obtained.

4.3.7 *Micrometer cylinder gauge*

Figure 4.21(b) show the principle of the micrometer cylinder gauge. It is used for measuring the diameters of holes to a high degree of accuracy. It uses a micrometer-controlled wedge to expand three equispaced anvils until they touch the walls of the bore. Unfortunately it only has a limited measuring range and the range cannot be extended by the use of extension rods (see 'Internal micrometer' above). A separate instrument has to be used for each range of hole sizes.

Fig. 4.21 *Further applications of the micrometer principles: (a) the internal micrometer; (b) the micrometer cylinder gauge; (c) micrometer depth gauge; (d) applications of the micrometer depth gauge*

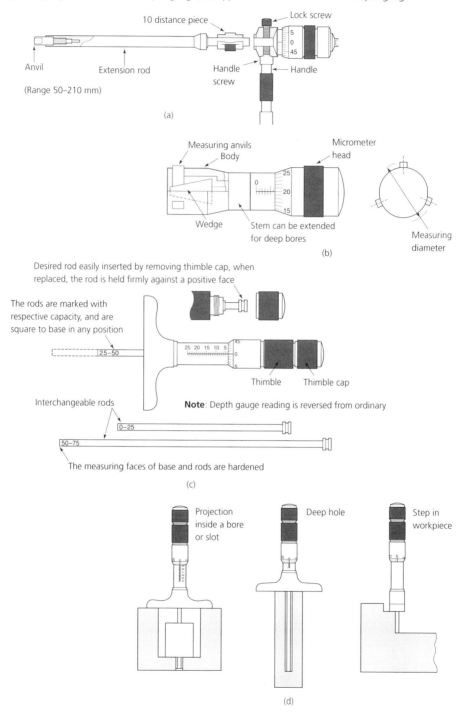

4.3.8 *Depth micrometer*

This is used for measuring the depth of holes and slots. You must take care when using a depth micrometer because its scales are reversed when compared with the familiar micrometer caliper. Also the measuring pressure tends to lift the micrometer off its seating. A depth micrometer is shown in Fig. 4.21(c). The measuring range is 25 mm for any given rod. Typical rods give a range of 0–25 mm, 25–50 mm, 50–75 mm. Some applications of a depth micrometer are shown in Fig. 4.21(d).

4.3.9 *Vernier calipers*

Although more cumbersome to use and rather more difficult to read, the vernier caliper has three main advantages over the micrometer caliper.

- One instrument can be used for measurements ranging over the full length of its main (beam) scale. Figure 4.22(a) shows a vernier caliper.
- It can be used for both internal and external measurements, as shown in Fig. 4.22(b). Remember that for internal measurements you have to add the combined thickness of the jaws to the scale readings.
- One instrument can be used for taking measurements in both inch units and in metric dimensional systems.

Fig. 4.22 *The vernier caliper: (a) construction; (b) use; (c) vernier scale (50 divisions)*

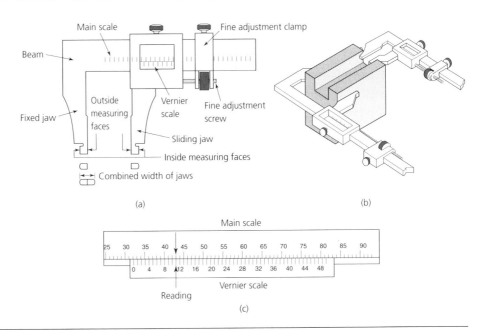

The measuring accuracy of a vernier caliper tends to be of a lower order than that obtainable with a micrometer caliper because:

- It is difficult to obtain a correct 'feel' with this instrument due to its size and weight.
- The scales can be difficult to read accurately even with a magnifying glass.

All vernier type instruments have two accurately engraved scales. A main scale marked in standard increments of measurement like a rule, and a vernier scale which slides along the main scale. This vernier scale is marked with divisions whose increments are slightly smaller than those of the main scale. Some vernier calipers are engraved with both inch and millimetre scales.

In the example shown in Fig. 4.22(c) the main scale is marked off in 1.00 mm increments, whilst the vernier scale has 50 divisions marked off in 0.98 mm increments. This enables you to read the instrument to an accuracy of $1.00 - 0.98 = 0.02$ mm. The reading is obtained as follows:

- Note how far the zero of the vernier scale has moved along the main scale (32 'whole' millimetres in this example).
- Note the vernier reading where the vernier and main scale divisions coincide (11 divisions in this example). You then multiply the 11 divisions by 0.02 mm which gives you 0.22 mm.
- Add these two readings together:

32 'whole' millimetres	$= 32.00$ mm *plus*
11 vernier divisions	$= 00.22$ mm
reading shown in Fig. 4.22(c)	$= 32.22$ mm

Always check the scales before use as there are other systems available and not all vernier scales have 50 increments. This is particularly the case in some cheap instruments. Also check that the instrument reads zero when the jaws are closed. If not, then the instrument has been strained and will not give a correct reading. There is no means of correcting this error and the instrument must be scrapped. Since they are expensive, it is essential to treat them with care. Figures 4.23 and 4.24 show, respectively, some additional vernier scales in metric and inch units for reading practice. Try to work them out for yourself before looking at the correct readings given at the end of this chapter (page 139).

Battery-operated direct reading digital calipers are also available, However, these tend to be heavier and more bulky and are often not as convenient to use as vernier calipers despite being easier to read. Digital micrometers are also available and the same comments apply to them. Digital electronic verniers and micrometers are very much more expensive than traditional instruments. As for all measuring instruments vernier calipers must be treated with care and cleaned before and after use. They should always be kept in the case provided. This not only protects the instrument from damage, it also supports the beam and prevents it from becoming distorted.

The vernier principle can also be applied to height gauges and depth gauges. Examples of these instruments are shown in Fig. 4.25. Some uses of the vernier height gauge are described later in this chapter and in Chapter 5.

Fig. 4.23 *Vernier scales (metric) reading exercises. (Note: The scales shown in all these exercises have a reading accuracy of 0.02 mm)*

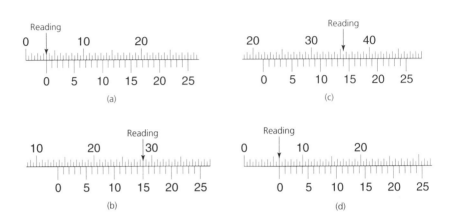

Fig. 4.24 *Vernier scales (English) reading exercises. (Note: The scales shown in all these exercises have a reading accuracy of 0.001 inch)*

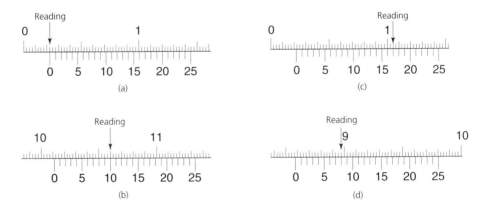

Fig. 4.25 *Vernier gauges: (a) height gauge; (b) depth gauge; (c) use of the depth gauge*

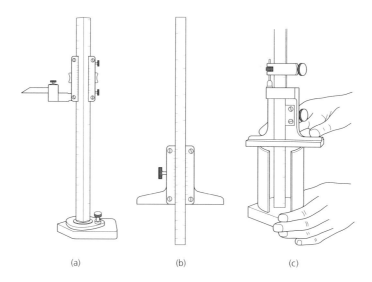

(a) (b) (c)

4.3.10 *Dial test indicators*

Dial test indicators (DTI) are often referred to as 'clocks' because of the appearance of the dial and pointer. They measure the displacement of a plunger or stylus and indicate the magnitude of the displacement on a dial by means of a rotating pointer. There are two main types of dial test indicator.

Plunger type

This type of instrument relies upon a rack-and-pinion mechanism to change the linear (straight line) movement of the plunger into rotary motion for the pointer. A gear train is used to magnify the movement of the pointer. This type of instrument has a long plunger movement and is, therefore, fitted with a secondary scale to count the number of revolutions made by the main pointer. A large range of dial diameters and markings are available. Figure 4.26(a) shows a typical example of this type of instrument and Fig. 4.26(b) shows how you can use one of these instruments to make comparative measurements.

Figure 4.27 shows how a dial test indicator can be used for setting up workpieces and aligning work-holding devices.

Fig. 4.26 *Dial test indicator (DTI): (a) plunger type; (b) comparative measurement*

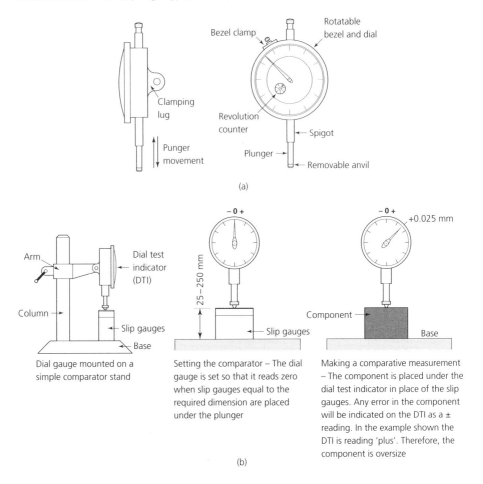

(a)

Dial gauge mounted on a simple comparator stand

Setting the comparator – The dial gauge is set so that it reads zero when slip gauges equal to the required dimension are placed under the plunger

Making a comparative measurement – The component is placed under the dial test indicator in place of the slip gauges. Any error in the component will be indicated on the DTI as a ± reading. In the example shown the DTI is reading 'plus'. Therefore, the component is oversize

(b)

Lever type

This type of instrument uses a lever and scroll to magnify the displacement of the stylus. Compared with the plunger type, the lever type instrument has only a limited range of movement. However, it is extremely popular for inspection and machine setting because it is more compact and the scale is more conveniently positioned for these applications. Figure 4.28(a) shows a typical example of this type of instrument and Fig. 4.28(b) shows an application of its use. In this example the DTI is mounted on a vernier height gauge, and ensures that the measuring, contact, pressure over H_1 and H_2 is constant. That is, in each position, the vernier height gauge is adjusted until the DTI reads zero before the height gauge reading is taken.

Fig. 4.27 *Machine setting and alignment using a dial test indicator: (a) setting work to run true on a lathe; (b) setting a milling machine vice; (c) setting up work on a rotary table; (d) checking surface for parallelism (constant reading of DTI indicates upper surface of component is parallel with machine table)*

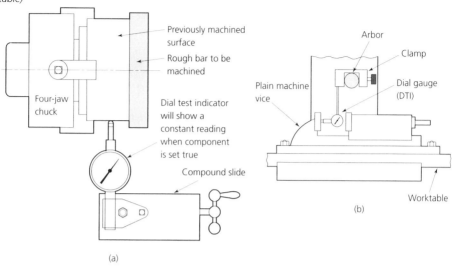

Previously machined surface

Rough bar to be machined

Four-jaw chuck

Dial test indicator will show a constant reading when component is set true

Compound slide

(a)

Arbor

Plain machine vice

Clamp

Dial gauge (DTI)

Worktable

(b)

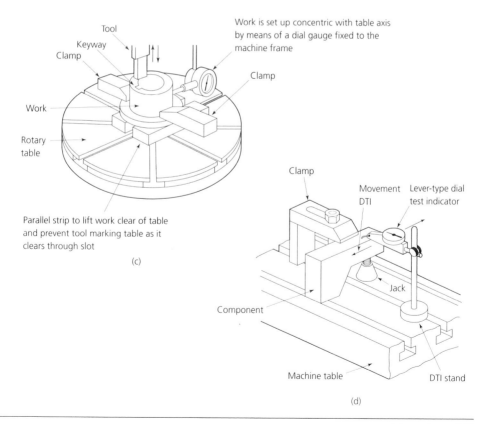

Tool

Keyway

Clamp

Work is set up concentric with table axis by means of a dial gauge fixed to the machine frame

Clamp

Work

Rotary table

Parallel strip to lift work clear of table and prevent tool marking table as it clears through slot

(c)

Clamp

Movement

DTI

Lever-type dial test indicator

Jack

Component

Machine table

DTI stand

(d)

Fig. 4.28 **Fig. 4.28** *Dial test indicator (DTI): (a) lever type; (b) use of a DTI to ensure uniform measuring pressure – by adjusting the vernier height gauge until the dial test indicator reads zero for each measurement taken, errors of 'feel' are removed*

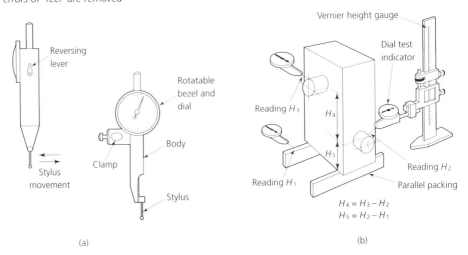

(a)

(b)

4.3.11 *Slip gauges*

Slip gauges are blocks of steel that have been hardened and stabilised by heat treatment. They are ground and lapped to size to very high standards of accuracy and surface finish. They are the most accurate standards of length available for use in workshops. The accuracy and finish is so high that two or more slip gauges may be *wrung* together. The method of wringing slip gauges together is shown in Fig. 4.29(a). When correctly cleaned and wrung together, the individual slip gauges adhere to each other by molecular attraction and, if left like this for too long, a partial cold weld will take place. If this is allowed to occur, the gauging surfaces will be irreparably damaged when the blocks are separated. Therefore, immediately after use, the gauges should be separated carefully, as shown in Fig. 4.29(b). They should then be cleaned, smeared with petroleum jelly (vaseline) and returned to their case. A typical 78-piece metric set of slip gauges is listed in Table 4.1.

Fig. 4.29 *Slip gauges: (a) 'wring' the gauges together with a rotary motion to assemble them; (b) slide the gauges to separate them. (Note: Pulling the slip gauges apart damages the gauging faces)*

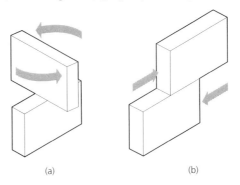

(a)

(b)

Table 4.1 Metric slip gauges

Range (mm)	Steps (mm)	Pieces
1.01 to 1.49	0.01	49
0.50 to 9.50	0.50	19
10.00 to 50.00	10.00	5
75.00 and 100.00	—	2
1.002 5	—	1
1.005	—	1
1.007 5	—	1

In addition, some sets also contain *protector slips* which are 2.50 mm thick and are made from a hard, wear-resistant material such as tungsten carbide. These are added to the ends of the slip gauge stack to protect the other gauge blocks from wear. Allowance must be made for the thickness of the protector slips when they are used.

Slip gauges are stacked together to give the required dimension. In order to achieve the maximum accuracy the following precautions must be preserved:

- Use the minimum number of blocks.
- Wipe the measuring faces clean using a soft clean 'chammy' leather.
- *Wring* the individual blocks together.

Let's see how we can build up a stack of slip gauges to give a dimension of 39.9725 mm. This is shown in Fig. 4.30 and four slips are the minimum we can use in this example.

- The first slip selected always gives the right-hand digit(s). In this case 1.0025 mm.
- The second slip and the third slip have been chosen to give the remaining decimal places (1.470 + 7.500 = 8.970 mm).

Fig. 4.30 *Building up slip gauges*

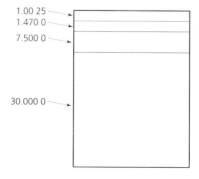

- The fourth slip provides the balance of the whole number $(39.000 - 1.0025 - 1.470 - 7.500 = 30\,mm)$. Thus the fourth slip is $30\,mm$. All these sizes are available in the set listed in Table 4.1.
- If protector slips had been used $(2 \times 2.5\,mm = 5.0\,mm)$ then the 7.500 slip would have been replaced by a $2.50\,mm$ slip.

Slip gauges come in various grades – workshop, inspection and standards room – and their accuracy and cost increases accordingly. They may be used directly for checking the width of slots. They may also be used in conjunction with a DTI for measuring heights and they may also be used for setting comparators, as was shown in Fig. 4.26.

Slip gauge accessories
Sets of slip gauge accessories, such as those shown in Fig. 4.31(a), are available for building up precision gauging and marking-out devices. Some examples are shown in Fig. 4.31(b). The examples shown indicate the convenient manner in which slip gauges and slip gauge accessories can be used to build up special measuring, marking out and limit gauging devices for single components and small production batches where the number of components does not warrant the use of specially manufactured gauges, or where an accuracy greater than normal is required.

Fig. 4.31 *Slip gauge accessories: (a) set of slip gauge accessories; (b) basic cage and type 'B' jaw; (c) small cage and pair of 'A' jaws; (d) large cage and pair of 'A' jaws; (e) small cage and pair of scribing points*

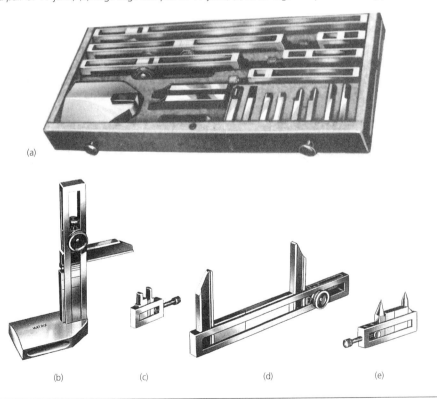

(a)

(b)   (c)   (d)   (e)

4.3.12 *Length bars*

For high-precision length measurements beyond the range of slip gauges, combination length bars are used. These come in a set, as shown in Fig. 4.32(a). Table 4.2 gives an indication of the range of sizes and accuracy of a typical set of workshop length bars.

You should note that on all bars of length 200 mm and over, two *Airey bands* are engraved equidistant from each end. When the bars are used in the horizontal position they must be supported at these points. They must also be supported at these points when not in use and their storage case is designed to provide such support. If the supports are too close together, *hogging* occurs, as shown in Fig. 4.32(b). If the supports are too far apart *sagging* occurs, as shown in Fig. 4.32(c). You can see that when hogging and sagging occur, the end faces of the length bars will not be parallel and measurement becomes inaccurate.

Fig. 4.32 *Combination length bars: (a) set of bars; (b) the supports are too close together – the bar 'hogs'; (c) the supports are too far apart – the bar 'sags'; (d) shows the correct support of length bars (the Airey points)*

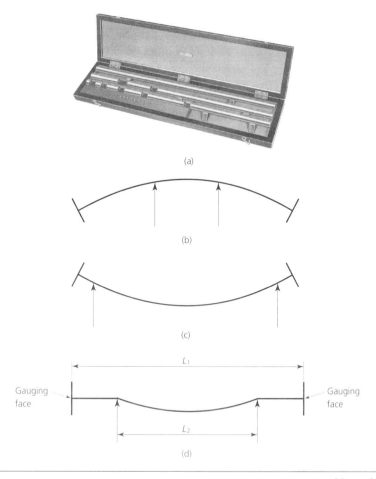

Table 4.2 **Range and accuracy of workshop grade metric length bars**

Nominal size (mm)	Flatness (mm)	Parallelism (mm)	Squareness to common axis of Airey bands over diameter of face* (mm)	Tolerance of size of 20 °C (mm)
10	0.000 20	0.000 25	—	+0.0003 −0.0003
20	0.000 20	0.000 25	—	+0.0003 −0.0003
40	0.000 20	0.000 25	—	+0.0004 −0.0003
60	0.000 25	0.000 25	—	+0.0006 −0.0003
80	0.000 25	0.000 40	—	+0.0008 −0.0004
100	0.000 25	0.000 40	—	+0.0010 −0.0005
200	0.000 25	0.000 40	0.001 25	+0.0020 −0.0010
300	0.000 25	0.000 40	0.001 25	+0.0030 −0.0015
400	0.000 25	0.000 40	0.001 25	+0.0040 −0.0020
600	0.000 25	0.000 40	0.001 25	+0.0060 −0.0030
800	0.000 25	0.000 40	0.001 25	+0.0080 −0.0040

* It is intended that bars 125 mm in length and shorter should be used unsupported, therefore no squareness figure is specified.

However, when the bars are supported at their *Airey points* (named after the physicist who investigated this problem), sagging is limited and only sufficient to pull the gauging faces up until they are parallel with each other and parallel to the measuring plane, as shown in Fig 4.32(d). There is a mathematical formula for calculating the *Airey points*, but all you need to know is that, for two points of support,

$$L_2 = 0.577 L_1$$

You may hear people talk about points of *minimum deflection* when supporting long straight edges. This is something quite different and not to be confused with the *Airey points*.

Length bars cannot be wrung together like slip gauges because of their weight and length. They are joined together by screwed dowels, as shown in Fig. 4.33(a). Figure 4.33(b) shows how slip gauges are used to obtain intermediate lengths. Standard accessories are also available for length bars to build up caliper gauges and height gauges, as shown in Fig. 4.34.

Fig. 4.33 *Joining length bars: (a) length bars are joined with dowel screws; (b) obtaining intermediate lengths using slip gauges*

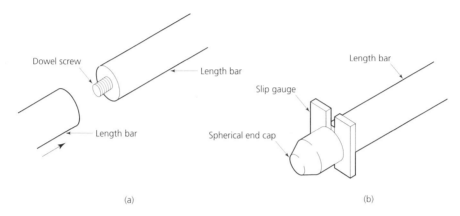

Fig. 4.34 *Some applications of length bars*

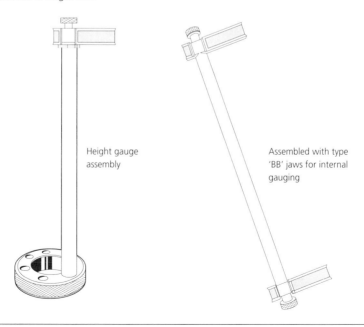

4.4 Measuring angles

Angles are measured in degrees and fractions of a degree. One degree of arc is 1/360 of a complete circle. One degree of arc can be subdivided into minutes and seconds (not to be confused with minutes and seconds of time):

60 seconds ($''$) of arc = 1 minute ($'$) of arc
60 minutes ($'$) of arc = 1 degree ($°$) of arc

With the introduction of calculators and computers, decimal fractions of a degree are also used. However, 1 minute of arc equals $0.016\,666\,6°$ recurring so there is no correlation between the two systems of subdividing a degree.

4.4.1 *Right angles*

A right angle is the angle between two surfaces that are at $90°$ to each other. Such surfaces may also be described as being *mutually perpendicular*. The use of engineers' try-squares and their use for scribing lines at right angles to the edge of a component will be described in Chapter 5. Figure 4.35(a) shows a typical engineer's try-square.

Note that a try-square is not a measuring instrument. It does not measure the angle. It only indicates whether or not the angle is a right angle. In Fig. 4.35(b), the stock is placed against the edge AB of the work and slid gently downwards until the blade comes into contact with the edge BC. Any lack of squareness will allow light to be seen between the edge BC and the try-square blade.

Fig. 4.35 *The try-square (a), its use (b) and how to check it (c)*

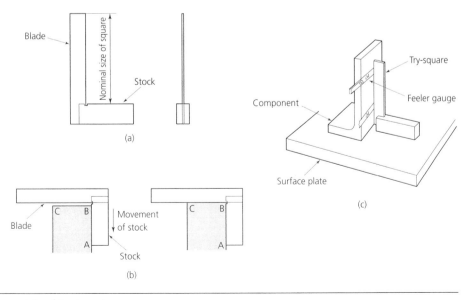

It is not always convenient to hold large work and a try-square up to the light. Figure 4.35(c) shows an alternative method using a surface plate as a datum surface. The squareness of the component face is checked with feeler gauges as shown. If the face is square to the base, the gap between it and the try-square blade will be constant.

Try-squares are precision instruments and should be treated with care if they are to retain their initial accuracy. They should be kept clean and lightly oiled after use. They should not be dropped, nor should they be kept in a drawer with other bench tools which may knock up burrs on the edges of the blade and stock. They should be checked for squareness at regular intervals. One simple technique is shown in Fig. 4.36. Two lines are drawn as shown. If the lines are not coincident (on top of each other) the square is faulty. Obviously, the square shown in Fig. 4.36 is faulty. Since the error is doubled by this check, the error is quite easy to see. More sophisticated equipment is required for a precise check.

Fig. 4.36 *Simple check for a try-square*

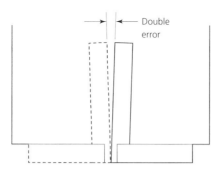

In addition to try-squares, prismatic squares and cylinder squares (Fig. 4.37) may be used for checking large work. Their design ensures that they make line contact with the work. Figure 4.38 shows a squareness comparator being used. In Fig. 4.38(a) the comparator is brought up to the square so that the fixed anvil makes contact with the

Fig. 4.37 *Cylinder square*

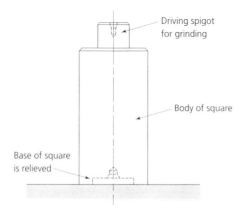

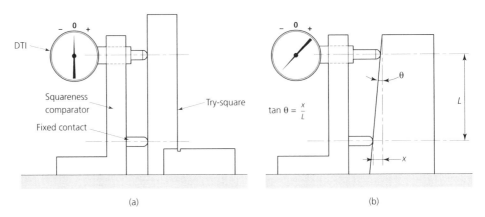

square. The DTI is then also in contact with the square and is set to read zero. In Fig.
4.38(b) the squareness comparator is in contact with the work. If the work is perpendicular
to the datum surface the DTI will again read zero. If not, then the reading of the DTI will
be the distance x in the figure. Knowing the distance L, the angular error can be calculated
by using trigonometry. The formula is given in Fig. 4.38(b). Try working it out for yourself
if $x = 0.10$ mm and $L = 100$ mm. I hope you have 'trig' ratios on your calculator. The
answer is given at the end of the chapter.

4.4.2 *Angles other than right angles*

Plain bevel protractor

Figure 4.39 shows a simple bevel protractor for measuring angles of any magnitude
between 0° and 180°. Such a protractor has only limited accuracy (±0.5°).

Fig. 4.39 *The plain bevel protractor (a) and its use in checking angles (b)*

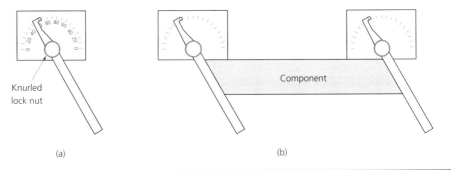

Vernier protractor

Where greater accuracy is required the vernier protractor should be used. The scales of a vernier protractor are shown in Fig. 4.40. The main scale is divided into degrees of arc, and the vernier scale has 12 divisions each side of zero. These vernier scale divisions are marked 0 to 60 minutes of arc, so that each division is 1/12 of 60, that is 5 minutes of arc. The reading for a vernier protractor is given by the sum of:

- The largest 'whole' degree on the main scale as indicated by the vernier zero mark.
- The reading of the vernier scale division in line with a main scale division.

Fig. 4.40 *Vernier protractor scales*

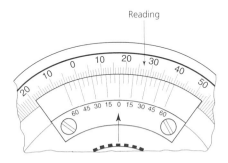

Thus the reading for the scales shown in Fig. 4.40 is:

17 'whole' degrees		$= 17°\ 00'$
vernier 25 mark in line with main scale	$=$	$00\ 25'$
Total angle	$=$	$17°\ 25'$

Vernier protractors are also available which can be read in degrees and decimal fractions of a degree.

4.4.3 *Sine bar*

Use of a sine bar is a simple but very accurate method of measuring and checking angles. Figure 4.41(a) shows a typical sine bar and Fig. 4.41(b) shows the principle of its use. The sine bar, slip gauges and the datum surface on which they stand form a right-angled triangle with the sine bar as the hypotenuse. Remember that the hypotenuse is the side opposite the right-angle in a right-angled triangle.

since $\quad \sin\theta = \dfrac{\text{opposite side}}{\text{hypotenuse}}$

then $\quad \sin\theta = \dfrac{\text{height of slip gauges}}{\text{nominal length of sine bar}}$

$\qquad\quad = \dfrac{H}{L}$

Fig. 4.41 *The sine-bar (a) and the principle of its use (b)*

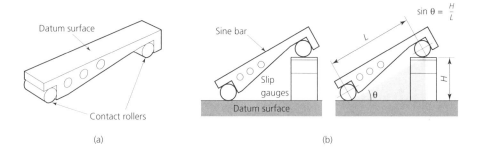

Datum surface

Contact rollers

(a)

Sine bar

Slip
gauges

Datum surface

(b)

$\sin \theta = \dfrac{H}{L}$

L

θ

H

Figure 4.42 shows a component being checked. The slip gauges are chosen to incline the sine bar at the required angle. The component is then placed on the sine bar. The height over the component is checked with a DTI mounted on a surface gauge or a vernier height gauge. If the component has been manufactured to the correct angle, then the DTI will read the same at each end of the component. Any difference in the readings indicates the magnitude of error in the angle of the component. Try working out the height of the slip gauges you would require to give an angle of $\theta = 25°$ if the nominal length (L) of the sine bar is 200 mm. Also state how you would build up the slip gauge stack. The answers are given at the end of the chapter.

Fig. 4.42 *Use of the sine bar*

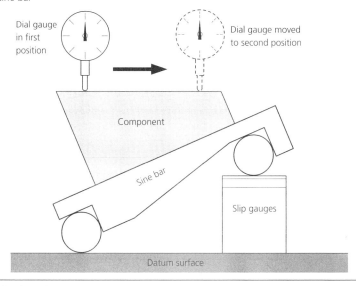

Dial gauge
in first
position

Dial gauge moved
to second position

Component

Sine bar

Slip gauges

Datum surface

4.4.4 *Taper plug and ring gauges*

Figure 4.43 shows typical taper plug and ring gauges. These cannot measure the angle of taper but they can indicate whether or not the taper is of the correct diameter. The gauges are 'stepped' as shown. If the component is within its 'limits of size' then the end of the taper will lie within the step.

Fig. 4.43   *Taper plug and ring gauges*

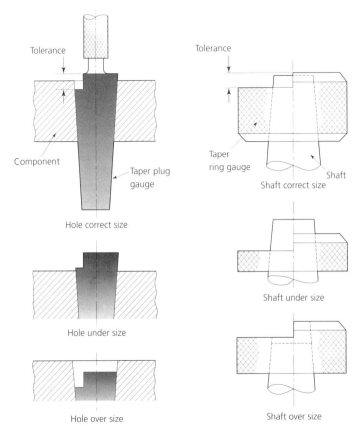

As already stated, taper plug and ring gauges cannot measure the angle of taper, but they can be used to check the angle, as shown in Fig. 4.44. Although a plug gauge is shown, a ring gauge can be used in a similar manner.

Plug gauge
- 'Blue' the gauge with a light smear of engineer's 'blue' and insert the gauge into the hole.
- Remove the gauge, taking care not to rock or rotate the plug gauge.
- Wipe the gauge clean of any remaining 'blue'.

- Re-insert the gauge carefully into the hole.
- Upon withdrawing the gauge, the smear left upon it will indicate the area of contact. This is interpreted as shown in Fig. 4.44.

Fig. 4.44 *Checking the angle of taper: (a) 'smear' indicates that the hole has the same taper as the plug gauge; (b) 'smear' indicates that the hole has a smaller angle of taper than the gauge; (c) 'smear' indicates that the hole has a larger angle of taper than the gauge*

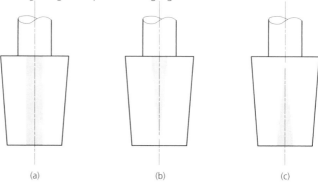

(a)   (b)   (c)

Ring gauge
- Lightly 'blue' the shaft and insert it carefully into the ring gauge.
- Remove the shaft and wipe it clean.
- Re-insert the shaft into the gauge.
- Withdraw the shaft and the smear will indicate the area of contact.
- Interpretation of the smear is similar to that when a plug gauge was used.

4.5 Dimensional deviation

The need for tolerance dimensions and the types of fit that can be obtained have already been introduced in Chapter 3 when dimensioning drawings. We must now consider these concepts as they apply to measurement. The upper and lower sizes of a dimension are called the *limits* and the difference in size between the limits is called the *tolerance*. The terms associated with limits and fits can be summarised as follows:

- *Nominal size* This is the dimension by which a feature is identified for convenience. For example, a slot of 25.15 mm actual width would be known as the 25 mm wide slot.
- *Basic size* This is the exact functional size from which the limits are derived by application of the necessary allowance and tolerances. The basic size and the nominal size are often the same.
- *Actual size* The measured size corrected to what it would be at 20 °C.
- *Limits* These are the high and low values of size between which the size of a component feature may lie. For example, if the lower limit of a hole is 25.05 mm and the upper limit of the same hole is 25.15 mm, then a hole which is 25.1 mm diameter is *within limits* and is acceptable.
- *Tolerance* This is the difference between the limits of size. That is, the upper limit minus the lower limit. Tolerances may be bilateral or unilateral, as shown in Fig. 4.45.

Fig. 4.45 *Types of tolerance: (a) unilateral tolerances – tolerance zone does not cross the basic size; (b) bilateral tolerances – tolerance zone always crosses the basic size*

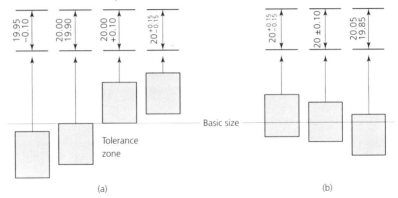

- *Deviation* This is the difference between the basic size and the limits. The deviation may be symmetrical, in which case the limits are equally spaced above and below the basic size. For example, 50.00 ± 0.15 mm. Alternatively, the deviation may be asymmetrical, in which case the deviation may be greater on one side of the basic size than on the other, e.g. 50.00 + 0.25 or −0.05.
- *Mean size* This size lies halfway between the upper and lower limits of size and must not be confused with either the nominal size or the basic size. It is only the same as the basic size when the deviation is symmetrical.
- *Minimum clearance (allowance)* This is the clearance between a shaft and a hole under maximum metal conditions – that is, the largest shaft in the smallest hole that the limits will allow. It is the tightest fit between shaft and hole that will function correctly. With a *clearance fit* the allowance is positive. With an *interference fit* the allowance is negative. These types of fit are discussed in Section 4.6.

4.6 Classes of fit

We have already seen the classes of fit that may be obtained between mating components in Fig. 3.20 (page 64). In the *hole basis system* the hole size is kept constant and the shaft size is varied to give the required class of fit. In an *interference fit* the shaft is always slightly larger than the hole. In a *clearance fit* the shaft is always slightly smaller than the hole. A *transition fit* occurs when the tolerances are so arranged that under maximum metal conditions (largest shaft; smallest hole) an interference fit is obtained, and that under minimum metal conditions (largest hole; smallest shaft) a clearance fit is obtained.

In a *shaft basis system* the shaft size is kept constant and the hole size is varied to give the required class of fit. Again, the classes of fit are: *interference fit, transition fit,* and *clearance fit.*

The hole basis system is the most widely used since most holes are produced by using standard tools such as drills and reamers. It is then easier to vary the size of the shaft by turning or grinding to give the required class of fit. Shaft and hole basis systems are shown in Fig. 4.46.

Tables of limits and fits and the instructions for their application to engineering components are set out in BS EN 20286-2 (previously BS 4500). A detailed study of BS EN 20286 is beyond the scope of this book, but some typical ISO fits (hole basis) have been included in Table 4.3. An example of the application of these standard limits and fits is shown in Fig. 4.47. Let's see how the system works.

Fig. 4.46 *Basis systems: (a) hole basis system; (b) shaft basis system*

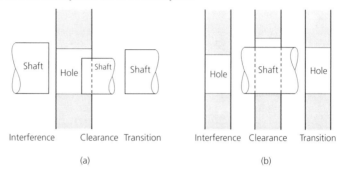

Table 4.3 *Some selected ISO fits – hole basis*

Normal sizes		Loose clearance		Average clearance		Close clearance		Precision clearance		Transition clearance		Interference	
Over (mm)	Up to (mm)	H9	e9	H8	f7	H7	g6	H7	h6	H7	k6	H7	p6
–	3	+25 / +0	−14 / −39	+14 / +0	−6 / −16	+10 / +0	−2 / −8	+10 / +0	−0 / −6	+10 / +0	+6 / +0	+10 / +0	+12 / +6
3	6	+30 / +0	−20 / −50	+18 / +0	−10 / −22	+12 / +0	−4 / −12	+12 / +0	−0 / −8	+12 / +0	+9 / +1	+12 / +0	+20 / +12
6	10	+36 / +0	−25 / −61	+22 / +0	−13 / −28	+15 / +0	−5 / −14	+15 / +0	−0 / −9	+15 / +0	+10 / +1	+15 / +0	+24 / +15
10	18	+43 / +0	−32 / −75	+27 / +0	−16 / −34	+18 / +0	−6 / −17	+18 / +0	−0 / −11	+18 / +0	+12 / +1	+18 / +0	+29 / +18
18	30	+52 / +0	−40 / −92	+33 / +0	−20 / −41	+21 / +0	−7 / −20	+21 / +0	−0 / −13	+21 / +0	+15 / +2	+21 / +0	+35 / +22
30	50	+62 / +0	−50 / −112	+39 / +0	−25 / −50	+25 / +0	−9 / −25	+25 / +0	−0 / −16	+25 / +0	+18 / +2	+25 / +0	+42 / +26
50	80	+74 / +0	−60 / −134	+46 / +0	−30 / −60	+30 / +0	−10 / −29	+30 / +0	−0 / −19	+30 / +0	+21 / +2	+30 / +0	+51 / +32
80	120	+87 / +0	−72 / −159	+54 / +0	−36 / −71	+35 / +0	−12 / −34	+35 / +0	−0 / −22	+35 / +0	+25 / +3	+35 / +0	+59 / +37
120	180	+100 / +0	−85 / −185	+63 / +0	−43 / −83	+40 / +0	−14 / −39	+40 / +0	−0 / −25	+40 / +0	+28 / +3	+40 / +0	+68 / +43
180	250	+115 / +0	−100 / −215	+72 / +0	−50 / −96	+46 / +0	−15 / −44	+46 / +0	−0 / −29	+46 / +0	+33 / +4	+46 / +0	+79 / +50
250	315	+130 / +0	−110 / −240	+81 / +0	−56 / −108	+52 / +0	−17 / −49	+52 / +0	−0 / −32	+52 / +0	+36 / +4	+52 / +0	+88 / +56
315	400	+140 / +0	−125 / −265	+89 / +0	−62 / −119	+57 / +0	−18 / −54	+57 / +0	−0 / −36	+57 / +0	+40 / +4	+57 / +0	+98 / +62
400	500	+155 / +0	−135 / −290	+97 / +0	−68 / −131	+63 / +0	−20 / −60	+63 / +0	−0 / −40	+63 / +0	+45 / +5	+63 / +0	+108 / +68

Source: Abstract from BS data sheet 4500A, derived from BS EN 20286-2.

Fig. 4.47 *Application of limits and fits: (a) tolerance specification (precision clearance fit); (b) dimensional limits to give precision clearance fit as derived from BS 4500: 1969*

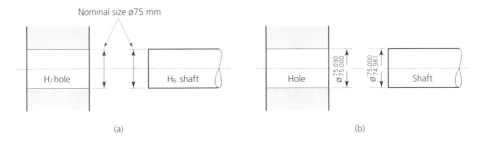

(a)

(b)

The tables provide for 28 types of shaft designated by lower-case letters, a, b, c, d, etc., and 28 types of hole designated by upper-case letters, A, B, C, D, etc. To each type of shaft or hole the grade of tolerance is designated by a number 01, 0, 1, 2, . . . , 16, thus giving 18 grades of tolerance in all. The letter indicates the position of the tolerance relative to the basic size and is called the *fundamental deviation*. The number indicates the magnitude of the tolerance and is called the *fundamental tolerance*. A shaft is completely defined by its basic size, letter and number, e.g. 75 mm h6. Similarly a hole is completely defined by its basic size, letter and number, e.g. 75 mm H7. Figure 4.48(a) shows how a precision clearance fit is specified using a 75 mm H7/h6 hole and shaft combination. Reference to Table 4.4 show that the hole dimensions will be:

$$75\,\text{mm} \quad \begin{array}{l} +0.030 \\ +0.000 \end{array}$$

and the shaft dimension will be:

$$75\,\text{mm} \quad \begin{array}{l} -0.000 \\ -0.019 \end{array}$$

Figure 4.48(b) shows how these dimensions are applied to the component drawing.

4.7 Accuracy

The greater the accuracy demanded by a designer, the narrower will be the tolerance band and the more difficult and costly it will be to manufacture the component within the limits specified. Table 4.4 shows the standard international tolerance (IT) bands. As the tolerance grade number (IT number) gets bigger, the tolerance gets bigger and the dimension becomes less precise and easier to achieve. You will also notice that as the nominal dimension gets bigger, the tolerance also gets bigger for any given IT number. This is because the larger a dimension becomes, the more difficult it becomes to hold any given level of accuracy. For example, a tolerance of 0.175 mm on a 2250 mm dimension has a greater accuracy than a tolerance 0.084 mm on a 22.5 mm dimension.

Table 4.4 Standard tolerance grades

Nominal sizes		Tolerance grades (µm)											Tolerance grades (mm)						
Over (mm)	Up to (mm)	IT1*	IT2*	IT3*	IT4*	IT5*	IT6	IT7	IT8	IT9	IT10	IT11	IT12	IT13	IT14†	IT15†	IT16†	IT17†	IT18†
—	3†	0.8	1.2	2	3	4	6	10	14	25	40	60	0.1	0.14	0.25	0.4	0.6	1	1.4
3	6	1	1.5	2.5	4	5	8	12	18	30	48	75	0.12	0.18	0.3	0.48	0.75	1.2	1.8
6	10	1	1.5	2.5	4	6	9	15	22	36	58	90	0.15	0.22	0.36	0.58	0.9	1.5	2.2
10	18	1.2	2	3	5	8	11	18	27	43	70	110	0.18	0.27	0.43	0.7	1.1	1.8	2.7
18	30	1.5	2.5	4	6	9	13	21	33	52	84	130	0.21	0.33	0.52	0.84	1.3	2.1	3.3
30	50	1.5	2.5	4	7	11	16	25	39	62	100	160	0.25	0.39	0.62	1	1.6	2.5	3.9
50	80	2	3	5	8	13	19	30	46	74	120	190	0.3	0.46	0.74	1.2	1.9	3	4.6
80	120	2.5	4	6	10	15	22	35	54	87	140	220	0.35	0.54	0.87	1.4	2.2	3.5	5.4
120	180	3.5	5	8	12	18	25	40	63	100	160	250	0.4	0.63	1	1.6	2.5	4	6.3
180	250	4.5	7	10	14	20	29	46	72	115	185	290	0.46	0.72	1.15	1.85	2.9	4.6	7.2
250	315	6	8	12	16	23	32	52	81	130	210	320	0.52	0.81	1.3	2.1	3.2	5.2	8.1
315	400	7	9	13	18	25	36	57	89	140	230	360	0.57	0.89	1.4	2.3	3.6	5.7	8.9
400	500	8	10	15	20	27	40	63	97	155	250	400	0.63	0.97	1.55	2.5	4	6.3	9.7
500	630*	9	11	16	22	32	44	70	110	175	280	440	0.7	1.1	1.75	2.8	4.4	7	11
630	800*	10	13	18	25	36	50	80	125	200	320	500	0.8	1.25	2	3.2	5	8	12.5
800	1000*	11	15	21	28	40	56	90	140	230	360	560	0.9	1.4	2.3	3.6	5.6	9	14
1000	1250*	13	18	24	33	47	66	105	165	260	420	660	1.05	1.65	2.6	4.2	6.6	10.5	16.5
1250	1600*	15	21	29	39	55	78	125	195	310	500	780	1.25	1.95	3.1	5	7.8	12.5	19.5
1600	2000*	18	25	35	46	65	92	150	230	370	600	920	1.5	2.3	3.7	6	9.2	15	23
2000	2500*	22	30	41	55	78	110	175	280	440	700	1100	1.75	2.8	4.4	7	11	17.5	28
2500	3150*	26	36	50	68	96	135	210	330	540	860	1350	2.1	3.3	5.4	8.6	13.5	21	33

* Values for standard tolerance grades IT1 to IT5 for basic sizes over 500 mm are included for experimental use.

† Standard tolerance grades IT14 to IT18 shall not be used for basic sizes less than or equal to 1 mm.

Note: Values for standard tolerance grades IT01 and IT0 for basic sizes less than or equal to 500 mm are given in ISO 286-1, Annex A, Table 5.

Source: BS EN 20286-2, previously BS 4500.

Therefore, for ease of manufacture at minimum cost, a designer never specifies an accuracy greater than is necessary to ensure the correct functioning of the component. Table 4.5 relates various manufacturing processes to the accuracy that can be expected from them under normal workshop conditions. The method of dimensioning can also affect the overall accuracy of a component. The effect of incremental (chain) dimensioning and absolute dimensioning (dimensioning from a common datum) on cumulative error was discussed in Chapter 3.

Table 4.5 *Process accuracy*

IT number	Class of work
16	Sand casting, flame cutting
15	Stamping
14	Die casting, plastic moulding
13	Presswork, and extrusion
12	Light presswork, tube drawing
11	Drilling, rough turning, boring
10	Milling, slotting, planing, rolling
9	Low grade capstan and automatic lathe work
8	Centre lathe, capstan and automatic
7	High quality turning, broaching, honing
6	Grinding, fine honing
5	Machine lapping, fine grinding
4	Gauge making, precision lapping
3	High-quality gap gauges
2	High-quality plug gauges
1	Slip gauges, reference gauges

4.8 Factors affecting accuracy

When measuring components, the more important factors affecting accuracy are given below.

4.8.1 *Temperature*

All metals and alloys expand when heated and contract when cooled. This is why measuring should take place in a constant temperature environment. You may have noticed that when you are machining materials in a workshop they often become hot. A component which has been heated by the cutting process will shrink whilst cooling to room temperature. As a result, a component that was within limits when measured on the machine may be found to be undersize when it is checked in the temperature-controlled inspection room. Take care.

Casting patterns and forging dies have to be made oversize to allow for shrinkage so that when the hot metal cools the components will not be undersize. The pattern maker usually works to a *contraction rule*. This has a normal scale on one side and an expanded scale on the other. By working to the expanded scale, the pattern maker automatically allows for the shrinkage that takes place as the casting metal solidifies and cools. Different rules have to be used for different casting metals, the expansion of the scale being matched to the shrinkage of a particular metal.

4.8.2 *Accuracy of equipment*

Since it is not possible to manufacture components to an exact size or to measure them to an exact size, it follows that measuring equipment cannot be made to an exact size. Therefore, measuring equipment also has to be manufactured to toleranced dimensions. In order that this has the minimum effect upon the measurement being made, *the accuracy of a measuring instrument should be about ten times greater than the accuracy of the component being measured.*

Measuring equipment should be checked regularly against even more accurate equipment. Where possible, any errors should be corrected by adjustment. If this is not possible, and the error has reached significant proportions, the instrument has to be discarded. In the case of standards such as slip gauges, these are checked against even more accurate standards. The actual deviation of size for each slip is charted. This is called *calibration* and it allows the cumulative error of a stack of slip gauges to be calculated. Allowance for this deviation can then be made when building up a stack of slip gauges or when setting a DTI to them.

4.8.3 *Reading errors*

There are two main reading errors:

- Misreading the instrument scales. Vernier scales are particularly difficult to read unless you have very good eyesight, so it is advisable to use a magnifying glass. Good lighting is also essential.
- Parallax (sighting errors) when using rules and similar scales. Care must be taken to ensure that your eye is over the point of measurement. The use of a solid abutment, as previously shown in Fig. 4.14, avoids you having to sight two points at the same time. Parallax or sighting errors can be minimised by using a rule which is as thin as possible. This was discussed in Section 4.2.

4.8.4 *Type of equipment*

It is possible to measure linear dimensions and angles with a variety of instruments. However, the accuracy of measurement is always lower than the reading accuracy and will depend, largely, upon the skill of the user. You must always match the instrument you use to the job in hand. It would be futile to try to measure an accurately machined dimension of 25.00 ± 0.02 mm with a rule and calipers. On the other hand, it would be a waste of time to use a vernier caliper when measuring a piece of bar in the stores to see if you could cut a 75 mm long blank from it. A rule would be quite adequate for this latter application.

4.8.5 *Line and end measurement*

This has been introduced earlier, but we will now look at it again in more detail.

Line measurement

As its name suggests, this is the measurement of the distance between two lines or two edges when measured with an instrument such as a steel rule where the distance to be measured is compared directly with an engraved scale. Sighting errors (parallax) make the measurement of the distance between two lines particularly difficult as datum abutments cannot, in this instance, be used. Dividers can be used to transfer the measurement. When setting the divider points to the lines, one leg of the dividers is 'clicked' into a main division on the rule and only the position on the rule scale of the second leg has to be sighted and read, the difference between the scale positions being the required measurement.

The measurement of the distance between an edge and a line is also difficult when using a line-measuring instrument such as a rule. The difficulties of such measurements can be eased by the use of a datum abutment, as previously shown in Fig. 4.14.

The difficulties associated with the measurement of the distance between two edges or across a diameter using a line-measuring instrument can be eased with the aid of calipers. The use of calipers has already been shown in Fig. 4.16.

End measurement

This is the preferred method of measurement for the engineer who is mostly concerned with the measurement of distances between faces, edges and across diameters. This measuring technique uses instruments such as micrometer calipers and vernier calipers where the edges or faces to be measured are in contact with the anvils or jaws of the measuring instrument and no sighting is required.

4.8.6 *Effect of force*

The use of excessive force when closing the measuring instrument on the workpiece being measured can cause distortion of both the workpiece and of the measuring instrument, resulting in an incorrect reading. In the worst case the distortion is permanent and either the workpiece or the measuring instrument, or both, can become worthless and have to be destroyed.

Some instruments are fitted with devices which ensure a correct and safe measuring pressure automatically. For example, three 'clicks' of the ratchet of a micrometer caliper applies the correct measuring force. The bench micrometer shown in Fig. 4.48 has a measuring force indicator (fiducial indicator) in place of the fixed anvil. When the pointers are in line, the correct measuring pressure is being applied. The force on the plunger of a DTI is limited by the strength of its return spring.

The contact area of the jaws or anvils of the measuring instrument can also influence the measuring pressure. This is because pressure is defined as force per unit area and, for any given measuring force, the contact pressure varies inversely with the contact area. Reduce the area and the measuring pressure is increased. Increase the contact area and the

Fig. 4.48 *The bench micrometer: The fiducial indicator removes errors of 'feel' – the micrometer is 'zeroed' with the pointer of the fiducial indicator in line with its datum mark and all subsequent measurements are made with the pointer in this position; this ensures constant measuring pressure*

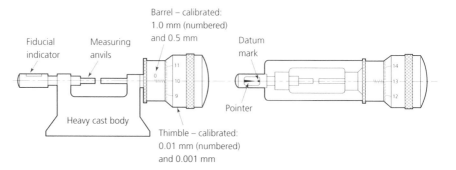

pressure is reduced. A spherically ended stylus will, in theory, result in point contact and this will give rise to an infinitely high measuring pressure. In practice the spherical end on the stylus tends to sink into the surface being measured, thus increasing the contact area. At the same time the spherical end of the stylus tends to flatten and this, again, increases the contact area. Any increase in the contact area results in a decrease in measuring pressure and a balance is automatically achieved between the measuring pressure and the resistance to deformation of the material of the component being measured. Such deformation introduces measuring errors and damage to the finished surfaces of the component being measured. Such effects are marginal where components are made from relatively hard metals but they must be taken into account when measuring components made from softer materials such as some plastics.

4.8.7 *Correct use of measuring equipment.*

No matter how accurately measuring equipment is made, and no matter how sensitive it is, one of the most important factors affecting the accuracy of measurement is the skill of the user. The more important procedures for the correct use of measuring equipment can be summarised as follows:

- The measurement must be made at right angles to the surface of the component.
- The use of a constant measuring pressure is essential. This is provided automatically with micrometer calipers by means of their ratchet. With other instruments such as plain calipers and vernier calipers the measuring pressure depends upon the skill and 'feel' of the user. Such skill only comes with practice and experience.
- The component must be supported so that it does not distort under the measuring pressure or under its own weight.
- The workpiece must be thoroughly cleaned before being measured, and coated with oil or a corrosion-inhibiting substance immediately after inspection. Ideally, gloves should be worn so that the acid in your perspiration does not corrode the cleaned surfaces of the instruments and the workpiece.

- Measuring instruments must be handled with care so that they are not damaged or strained. They must be cleaned and kept in their cases when not in use. Their bright surfaces should be lightly smeared with petroleum jelly (vaseline). Measuring instruments must be regularly checked to ensure that they have not lost their initial accuracy. If an error is detected the instrument must be taken out of service immediately so that the error can be corrected. If correction is not possible the instrument must be immediately discarded.

4.9 Terminology of measurement

Indicated size

This is the size indicated by the scales of a measuring instrument when it is being used to measure a workpiece. The indicated size makes no allowance for any incorrect use of the instrument, such as the application of excessive contact pressure.

Reading

This is the size as read off the instrument scales by the operator. Errors can occur by the user misreading the scales – for example, sighting (parallax) errors can occur when measuring with a rule. Vernier scales are particularly easy to misread in poor light. A magnifying lens is helpful in good light and even if you have good eyesight. Electronic measuring instruments with digital readouts overcome many of these reading difficulties.

Reading value

This is also called the 'reading accuracy'. This is the smallest increment of size that can be read directly from the scales of the instrument and will depend upon the layout of the scales. For example, micrometer calipers normally have a reading value of 0.01 mm, a bench micrometer fitted with a fiducial indicator will have a reading value of 0.001 mm, a vernier caliper with a 50 division vernier scale will have a reading value of either 0.01 mm or 0.02 mm, depending upon how the scales are arranged.

Measuring range

This is the range of sizes which can be measured by any given instrument. It is the arithmetical difference between the largest and smallest size that can be measured. For example, a 50–75 mm micrometer has a measuring range of $75 - 50 = 25$ mm.

Measuring accuracy

This is the actual accuracy expected from a measuring instrument after taking into account all the normal errors of usage. It can never be better than the indicated size.

4.10 General rules for accurate measurement

This chapter has covered many techniques of measurement and has recommended many procedures to ensure accurate measurement. The general rules for accurate measurement may be summarised as follows:

- Measuring equipment should be between two and ten times as accurate as the dimension being measured (the latter figure being achieved wherever possible). The same rule applies to gauges for checking dimensions.
- Accuracy is improved if the measured dimension is as close to the mean dimension as possible.
- Accuracy is improved if the size of the setting master used in comparative measurement is as close as possible to the size of the dimension. This limits the movement of the DTI and limits any cumulative error that may be present in the instrument.
- Wherever possible the measurement should be taken at the standard temperature of 20 °C.
- Measuring equipment must be carefully and correctly used. Excessive measuring forces must be avoided so that the equipment is not strained or the work distorted.
- Measuring equipment must be cleaned before and after use and wiped over with a corrosion preventative before being put away.
- The equipment must be inspected for errors and damage before and after use and there should be a regular programme of maintenance and recalibration.
- Measuring equipment must never be mixed up or stored with other bench tools and cutting tools. After use each item of measuring equipment must be returned to its case. The maker's recommendations for maintaining the equipment in good condition must always be observed.

EXERCISES

4.1 *Limits of size*
   (a) Figure 4.49 show a roller and its shaft. Complete the associated table from the dimensions given.
   (b) The limits of size for the width of a component are 11.5 and 12.5 mm. With the aid of sketches show **three** ways in which these limits of size may be applied to the dimension.
   (c) With the aid of sketches, show what is meant by:
       (i) a clearance fit
       (ii) a transition fit
       (iii) an interference fit
   (d) Which of the classes of fit listed in exercise (c) would be required for:
       (i) a pulley that is free to run on its shaft
       (ii) a drill bush that has to be pressed into the bushplate of a drilling jig

Fig. 4.49 *Exercise 4.1(a)*

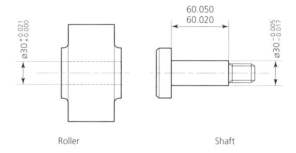

Roller

Shaft

	Roller hole diameter	Shaft diameter	Shaft diameter
Nominal size			
High limit			
Low limit			
Tolerance			
Minimum clearance (allowance) between roller and shaft			

4.2 *Measuring tools*

(a) List the most important features of an engineer's rule. State briefly how it should be cared for to maintain its accuracy.

(b) Explain briefly what are the main causes of reading error when using a steel rule and how these errors can be minimised.

(c) With the aid of sketches show the purpose of the following measuring tools and how they are used:
   (i) firm-joint calipers
   (ii) odd-leg (jenny) calipers

(d) With the aid of sketches show how an engineer's try-square is used to check the squareness of a rectangular blank that has been machined to 100 mm × 50 mm × 25 mm.

4.3 *Measuring instruments*

(a) Sketch a micrometer caliper and:
   (i) name its more important features
   (ii) show how the scales are arranged for metric readings
   (iii) show how the scales are arranged for 'inch' readings

(b) With the aid of sketches, explain how the scales of a depth micrometer differ from a micrometer caliper.

(c) Write down the micrometer readings shown in Fig. 4.50.

(d) Sketch a vernier caliper that can be used for internal and external measurements, and also depth measurements. Name its more important features.

(e) Write down the vernier readings shown in Fig. 4.51.

(f) With the aid of sketches, show how a plain bevel protractor is used to measure angles.

Fig. 4.50 *Exercise 4.3(c)*

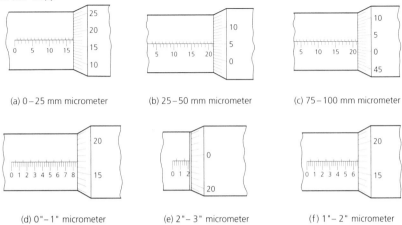

(a) 0 – 25 mm micrometer

(b) 25 – 50 mm micrometer

(c) 75 – 100 mm micrometer

(d) 0" – 1" micrometer

(e) 2" – 3" micrometer

(f) 1" – 2" micrometer

Fig. 4.51 *Exercise 4.3(e)*

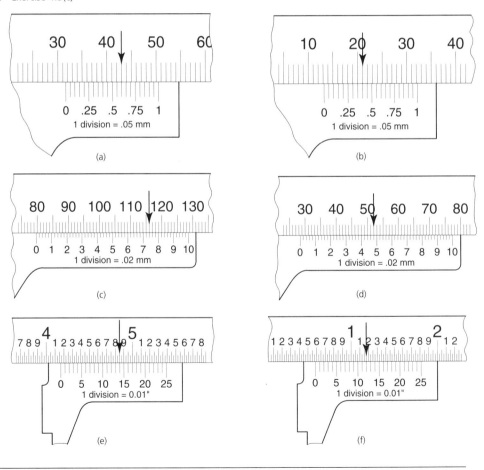

(a)

1 division = .05 mm

(b)

1 division = .05 mm

(c)

1 division = .02 mm

(d)

1 division = .02 mm

(e)

1 division = 0.01"

(f)

1 division = 0.01"

4.4 *Gauge blocks and length bars*

(a) Using the slip gauges (gauge blocks) listed in Table 4.1 (see p. 115), select a suitable set of gauge blocks to make up the dimension of 34.147 mm:
  (i) without using protector slips
  (ii) using protector slips

(b) With the aid of a sketch explain how slip gauges should be assembled together, and also taken apart, to avoid damage to the gauging surfaces.

(c) Explain how length bars are supported so that their measuring faces are kept parallel. Calculate the distance apart for the supports of a 1 metre long length bar.

(d) With the aid of sketches explain how slip gauges are used in conjunction with a sine bar to check a tapered component.

(e) Using slip gauges, a lever type (Verdict) dial test indicator (DTI) mounted on a scribing block, and a surface plate as a datum, describe with the aid of sketches, how the component shown in Fig. 4.52 can be checked for thickness and parallelism.

Fig. 4.52 *Exercise 4.4(e). (Dimensions in millimetres)*

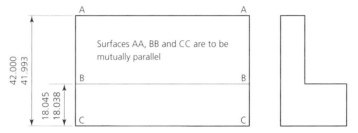

4.5 *Miscellaneous measuring devices*

(a) With the aid of sketches show how the following measuring devices are used:
  (i) radius gauges
  (ii) feeler gauges
  (iii) taper plug gauges (stepped)

ANSWERS

Fig. 4.20		**Fig. 4.23**	**Fig. 4.24**
(a) 0.178 inch	(f) 2.00 mm	(a) 3.50mm	(a) 0.225 inch
(b) 0.044 inch	(g) 5.23 mm	(b) 13.80 mm	(b) 10.110 inch
(c) 0.215 inch	(h) 17.78 mm	(c) 21.78 mm	(c) 0.217 inch
(d) 0.175 inch	(i) 0.31 mm	(d) 6.00 mm	(d) 8.583 inch
(e) 0.487 inch	(j) 6.05 mm		

5 Marking out

When you have read this chapter, you should be able to understand how to:
- Identify and select marking-out tools for making lines.
- Identify and select marking-out equipment for providing guidance.
- Identify and select marking-out equipment for providing support.
- Identify and select different types of datum.
- Identify and use different coordinate systems.
- Mark out workpieces having square, rectangular, circular and irregular-shaped sections.

5.1 Marking-out equipment (tools for making lines)

Marking out is, essentially, drawing on metal so as to provide guide lines for a fitter or a machinist to work to. A pencil line would not be suitable. The hard metal surface would soon make a pencil blunt and the line would become thick and inaccurate; also a pencil line is too easily wiped off a metal surface. Usually the line is scribed using a sharp-pointed metal tool that cuts into the surface of the metal and leaves a fine, permanent line. We will start by reviewing the more essential items of marking-out equipment.

5.1.1 *Scriber*

This is the basic marking-out tool. It consists of a handle with a sharp point. The pointed end is made from hardened steel so that it will stay sharp in use. Engineers' scribers usually have one straight end and one hooked end, as shown in Fig. 5.1. It is essential that the scribing point is kept sharp. Scribing points should not be sharpened on a grinding machine. The heat generated by this process tends to soften the point of the scriber so that it soon becomes blunt. The scribing point should be kept needle sharp by the use of an oil stone (see Section 5.7, Fig. 5.27).

Fig. 5.1 *Scriber*

5.1.2 *Centre and dot punches*

Typical punches are shown in Fig. 5.2. They are used for making an indentation in the surface of the metal. There are two types of punch. Figure 5.2(a) shows a dot punch. This has a relatively fine point of about 60° or less and is used for locating the legs of such instruments as dividers and trammels.

Figure 5.2(b) shows a centre punch. This is heavier than a dot punch and has a less acute point (usually 90° or greater). It is used to make a heavy indentation suitable for locating the point of a twist drill.

Another use for a dot punch is for 'preserving' a scribed line, as shown in Fig. 5.2(c). This use will be considered in greater detail towards the end of this chapter.

Fig. 5.2 *Punching: (a) dot punch; (b) centre punch; (c) protecting a scribed line*

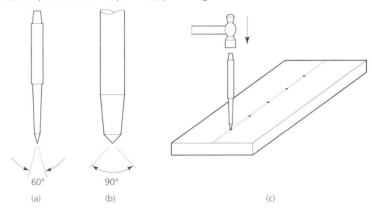

60°
(a)

90°
(b)

(c)

The correct way to using a dot punch is shown in Fig. 5.3. Usually the position for making a dot mark is at the junction of a pair of scribed lines at right angles to each other.

- You hold the punch so that it is inclined away from you. This enables you to see when the point of the punch is at the junction of the scribed lines, as shown in Fig. 5.3(a).
- You then carefully bring the punch up to the vertical, taking care not to move the position of the point.
- You then strike the punch lightly and squarely with a hammer, as shown in Fig. 5.3(b).
- Check the position of the dot with the aid of a magnifying glass. Draw the dot over if it is slightly out of position.

For rough work you can use a centre punch in the same way but you need to hit it harder with a heavier hammer if you are to make a big enough indentation to guide the point of a drill. Because of the difficulty in seeing the point of a centre punch it is preferable to make a dot punch mark and, when you are satisfied that it is correctly positioned, you can enlarge the dot mark with a centre punch. The centre punch is correctly positioned when you feel its point 'click' into the mark left by the dot punch.

Figure 5.3(c) shows an automatic dot punch. This has the advantage that it can be used single handed and it is less likely to skid across the surface of the work. The punch is operated by downward pressure that releases a spring-loaded hammer in its body. No separate hammer is required.

Fig. 5.3 *Correct way to use dot and centre punches*

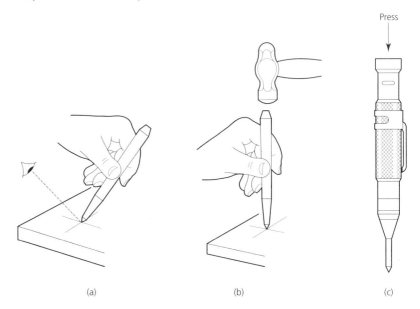

(a)   (b)   (c)

5.1.3 *Dividers and trammels*

These instruments are used for marking out circles and arcs of a circle. A typical pair of dividers and the names of its component parts are shown in Fig. 5.4(a). Dividers are used to scribe circular lines, as shown in Fig. 5.4(b). They are set to the required radius, as shown in Fig. 5.4(c). They are also used for stepping off equal distances (such as hole centres along a line or round a pitch circle) as shown in Fig. 5.4(d). The leg about which the dividers pivot is usually located in a fine centre dot mark. To locate the point of this leg accurately it is essential to use a sharp dot punch, as shown in Fig. 5.4(e).

Trammels are used for scribing large diameter circles and arcs that are beyond the range of ordinary dividers. They are also called beam compasses when the scribing points are located on a wooden beam, as shown in Fig. 5.4(f). Trammels have a metal beam usually in the form of a solid rod or a tube. This often carries a scale and one of the scribing points is fitted with a vernier scale and a fine adjustment screw for accurate setting, as shown in Fig. 5.4(g).

Fig. 5.4 *Dividers and trammels: (a) parts of a divider; (b) scribing a circle; (c) setting a required radius; (d) stepping off hole centres; (e) location of divider point; (f) trammel or beam compass; (g) adjustment of trammel*

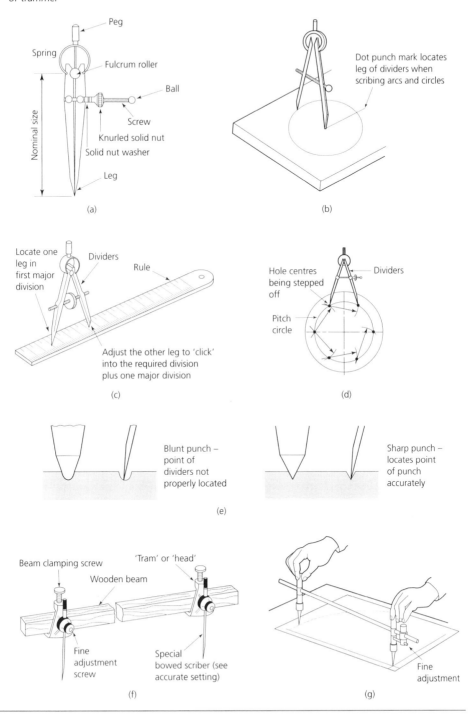

5.1.4 *Hermaphrodite calipers*

These are usually called odd-leg calipers or jenny calipers. They consist of one caliper leg and one divider leg and are used for scribing lines parallel to an edge, as shown in Fig. 5.5(a). They are set to the required size as shown in Fig. 5.5(b). Odd-leg calipers can also be used for finding the centre of round section bars and casting bosses, as shown in Fig. 5.5(c). The odd-legs are set to roughly half the diameter of the boss. Four lines are scribed at intervals of 120° around the boss. The centre of the boss is at the centre of the four intersecting lines as shown.

Fig. 5.5 *Hermaphrodite (odd-leg) calipers: (a) scribing lines parallel to an edge; (b) setting odd-leg calipers; (c) finding the centre of a bar*

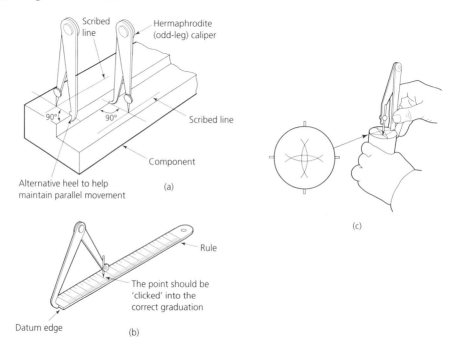

5.1.5 *Scribing block*

The scribing block or surface gauge is used for marking out lines parallel to a datum surface or a datum edge. The parts of a typical scribing block are shown in Fig. 5.6(a) and some typical applications are shown in Fig. 5.6(b). Normally the scribing point is set to mark a line at a given height above the base of the instrument. This line will be marked parallel to the surface along which the base of the instrument is moved. When a line parallel to a datum edge is required, the edge pins are lowered. These pins are then kept in contact with the edge as the scribing block is moved along the work.

Fig. 5.6 *The surface gauge (a) and typical applications (b)*

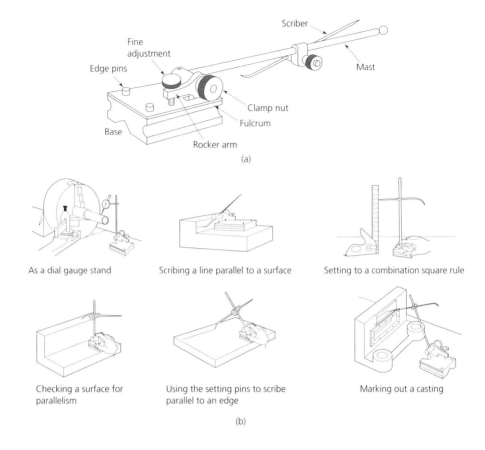

As a dial gauge stand

Scribing a line parallel to a surface

Setting to a combination square rule

Checking a surface for parallelism

Using the setting pins to scribe parallel to an edge

Marking out a casting

(b)

5.1.6 *Vernier height gauge*

The vernier height gauge was introduced in Chapter 4 as a measuring instrument. It is also used for scribing lines parallel to a datum surface in a similar manner to a scribing block. However, unlike a scribing block that has to be set to a separate steel rule, a vernier height gauge has a built-in main scale and vernier scale so that it can be set to a high degree of accuracy. The setting and reading of vernier scales was described in Chapter 4. The height gauge is fitted with a removable, sharpened nib. This is set to the required height by the scales provided. To scribe a line parallel to the datum surface, as shown in Fig. 5.7, the following procedure is used.

- Set the nib of the height gauge to the correct distance from the base of the instrument.
- Keep the base of the height gauge firmly on the datum surface on which it and the work are standing.
- Keep the scribing nib firmly in contact with the work surface.

- Move the height gauge across the datum surface so that the scribing nib slides across the work. Keep the nib at an angle to the work surface so that the nib trails the direction of movement.
- To sharpen the nib without losing the zero setting of the instrument, see Section 5.7, Fig. 5.27(b).

Fig. 5.7 *Use of a vernier height gauge to scribe a line parallel to a datum surface*

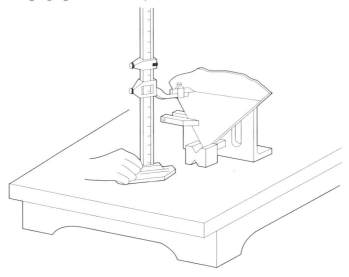

5.1.7 *Slip gauges*

Slip gauges are very accurate aids to fine measurement and were introduced in Chapter 4. They can be used with a variety of accessories to make up a range of marking-out and gauging devices and some examples were shown in Fig. 4.31. You will rarely have to work to this level of accuracy until you are fully skilled. Such equipment must be handled with great care and is mostly kept for inspection purposes.

5.2 Marking-out equipment (tools for providing guidance)

You cannot draw a straight line with a scriber without the help of some form of straight edge to guide the scriber.

5.2.1 *Rule and straight edge*

Where a straight line is required between two points, a rule can be used or, for longer distances, a straight edge. The correct way to use a scriber is shown in Fig. 5.8(a). The scriber is always inclined away from any guidance edge, such as a rule, straight edge or try-square. Its point should always trail the direction of movement to prevent it 'digging-in' to the metal surface; producing a poor line and damaging the scribing point.

5.2.2 *Box square*

This is also known as a key seat rule. It is used for marking and measuring lines scribed parallel to the axis of a cylindrical component such as a shaft. A typical box square and its method of use are shown in Fig. 5.8(b).

Fig. 5.8 *Scribing straight lines: (a) scribing a straight line using a rule or a straight edge; (b) scribing a straight line using a box square*

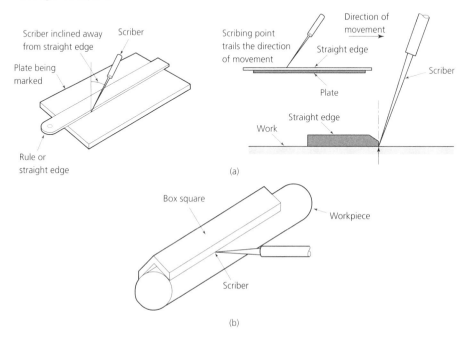

(a)

(b)

5.2.3 *Try-square*

When you need to scribe a line at 90° to an edge a try-square is used, as shown in Fig. 5.9. A line scribed at 90° to an edge or another line is said to be at *right angles* to that edge or line or is said to be *perpendicular* to that edge or line. They both mean the same thing.

Fig. 5.9 *Scribing a line perpendicular to an edge*

5.2.4 *Combination set*

This is shown in Fig. 5.10(a). It consists of a strong and relatively thick and rigid rule together with three 'heads' which are used singly but in conjunction with the rule.

- The square head can be clamped to the rule at any point along its length. It can either be used as a try-square (90°) or as a mitre square (45°), as shown in Fig. 5.10(b).
- The centre head or centre finder can also be clamped to the rule at any point along its length. The edge of the blade that passes through the centre of the centre finder also passes through the centre of the cylindrical workpiece. The centre of the cylindrical workpiece is found by scribing two lines at right angles to each other, as shown in Fig. 5.10(c). The lines intersect at the centre of the workpiece.
- A protractor head is also supplied and this is used for marking out lines that are at any angle other than 90° or 45° to the datum surface or edge.
- The square head and the protractor head are supplied with spirit (bubble) levels for setting purposes. However, they are only of limited accuracy.

5.3 Marking-out equipment (tools for providing support)

When marking out a component, it is essential that the blank is properly supported. As well as keeping the workpiece rigid and in the correct position, the supporting surface may also provide a datum from which to work. A datum is simply a line, surface or edge from which measurements are taken, but this will be discussed more fully in Section 5.4.

5.3.1 *Surface plate and tables*

Surface plates are cast from a stable cast iron alloy and are heavily ribbed to make them rigid. An example is shown in Fig. 5.11(a). They are used on the bench to provide a flat surface for marking out small workpieces. They are very heavy and should only be moved with care, preferably by two or more persons in the larger sizes.

Surface tables (marking-out tables), such as the one shown in Fig. 5.11(b), are used for providing a support and datum surface when marking out larger workpieces. A marking-out table is of heavy and rigid construction. The working surface may be of cast iron, machined or ground flat. Plate glass and granite are also used owing to their smoothness and stability, but they do not give such a nice 'feel' as cast iron when moving the instruments upon them for the simple reason that cast iron is self-lubricating.

The working surface must be kept clean and in good condition. Nothing must be allowed to scratch or damage the table and heavy objects must be slid gently onto the table from the side. Clean the table before and after use and make sure all sharp corners and rough edges are removed from the workpiece before it is placed on the table. Keep the table covered when it is not in use. Oil the working surface of the table if it is not to be used for some time.

Fig. 5.10 *The combination set: (a) construction; (b) uses; (c) finding the centre of a circular component*

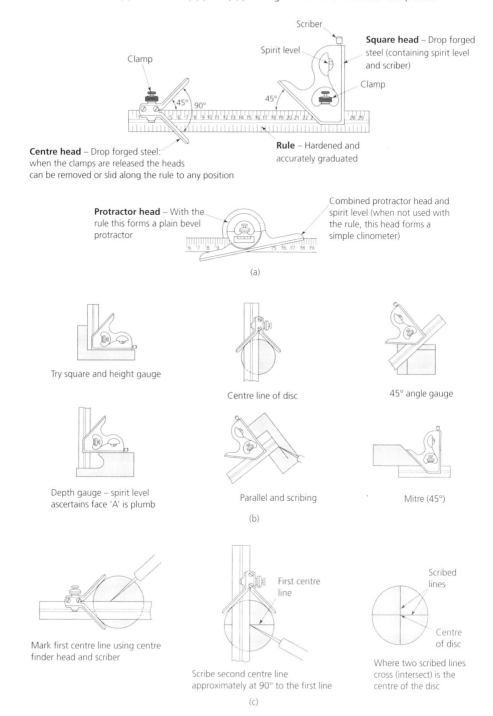

Scriber

Spirit level

Square head – Drop forged steel (containing spirit level and scriber)

Clamp

Clamp

45°

45°

90°

Centre head – Drop forged steel: when the clamps are released the heads can be removed or slid along the rule to any position

Rule – Hardened and accurately graduated

Protractor head – With the rule this forms a plain bevel protractor

Combined protractor head and spirit level (when not used with the rule, this head forms a simple clinometer)

(a)

Try square and height gauge

Centre line of disc

45° angle gauge

Depth gauge – spirit level ascertains face 'A' is plumb

Parallel and scribing

Mitre (45°)

(b)

Mark first centre line using centre finder head and scriber

First centre line

Scribe second centre line approximately at 90° to the first line

Scribed lines

Centre of disc

Where two scribed lines cross (intersect) is the centre of the disc

(c)

Fig. 5.11 *Surface plate (a) and marking out table (b)*

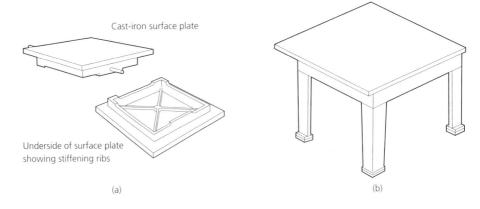

(a)   (b)

5.3.2 *Angle plates*

These are also made from good-quality cast iron and the working faces are machined at right angles to each other. The ends are also machined so that the angle plate can be stood on end when it is necessary to turn the work clamped to it through 90°. Figure 5.12(a) shows a typical angle plate being used to support work perpendicular to the datum surface of a marking-out table.

Figure 5.12(b) shows an adjustable angle plate. It is used for supporting work at any angle other than 90° to the datum surface of a marking-out table. There is usually a scale that can be used for initial setting. It is only of limited accuracy and a vernier protractor should be used for accurate setting.

Fig. 5.12 *Angle plates (a); (b) adjustable angle plate*

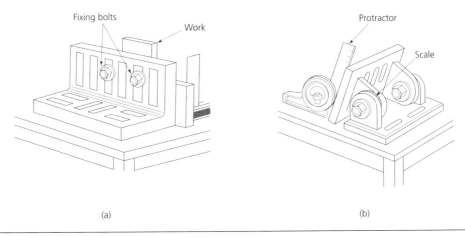

(a)   (b)

5.3.3 *Vee-blocks*

Vee-blocks are used for supporting cylindrical workpieces so that their axes (plural of axis) are parallel to the datum surface. They also prevent the work from rolling about. Figure 5.13(a) shows a pair of plain vee-blocks and Fig. 5.13(b) shows a pair of slotted vee-blocks with 'horseshoe' clamps. Vee-blocks are always manufactured as a matched pair and should be kept as a matched pair. This ensures the axis of the work is parallel to the datum surface of the marking-out table. Figure 5.13(c) shows some applications of vee-blocks. As well as vee-blocks, linked rollers are also used for supporting cylindrical work, as shown in Fig. 5.13(d).

Fig. 5.13 *Vee-blocks and linked rollers: (a) plain vee-blocks; (b) slotted vee-blocks with 'horseshoe' clamp; (c) uses of vee-blocks; (d) use of linked rollers*

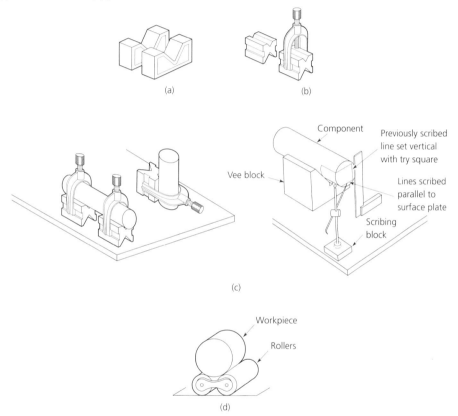

5.3.4 *Parallels*

These are parallel strips of hardened and ground steel of square or rectangular section. They are used for supporting and raising work. They are manufactured in various sizes and, like vee-blocks, are always manufactured in pairs. This ensures that the supported work is always parallel to the datum surface of the marking-out table.

5.3.5 *Jacks, wedges and shims*

Adjustable screw jacks are used to provide additional support for heavy castings, as shown in Fig. 5.14. Without the jack, the overhanging weight of the casting would make it unstable and it would tend to fall over.

Wedges are also useful in levelling heavy components, as shown in Fig. 5.14. Where wedges are too thick, shims can be used. Shims are cut from thin, hard-rolled brass or steel strip. The strip is supplied in graded thicknesses. They are used for packing the work level. It is always better to use one thick shim than two or more thin shims.

Fig. 5.14 *Supporting larger work*

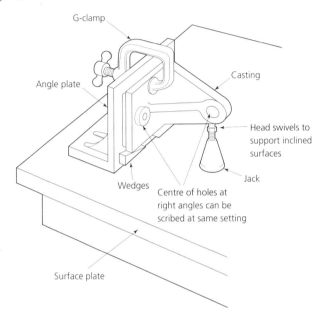

5.4 The advantages and disadvantages of manual marking out

For most jobbing work, prototype work, toolroom work and small quantity production, components are usually marked out as a guide to manufacture. The purposes, advantages and disadvantages of manual marking out can be summarised as follows.

5.4.1 *Purposes and advantages*

- To provide guide lines which are worked to, and which provide the only control for the size and shape of the finished component. This is only suitable for work of relatively low accuracy.

- To indicate the outline of the component to a machinist as an aid for setting up and roughing out. The final dimensional control would come, in this instance, from precision instruments used in conjunction with the micrometer dials of the machine itself.
- To ensure that adequate machining allowances have been left on castings and forgings before expensive machining operations commence. The features checked are surfaces, webs, flanges, cored holes and bosses. In the example shown in Fig. 5.15, it is obvious that the base will not clean up. The web is not central, nor will the bored hole be central in the boss. There would be no point in machining this casting.

5.4.2 *Disadvantages*

- Scribed lines cut into the surface of the workpiece and deface the surface of the metal. Where the surface finish is important, allowance must be left for surface grinding to remove the scribed marks on completion of the component. Any marks cut into the surface of the metal are a potential source of fatigue failure and cracking during heat treatment and bending.
- The above disadvantages cannot be overcome by drawing with a pencil as this would be neither sufficiently permanent nor sufficiently accurate. The only exception is in sheet metal work where fold lines are drawn with a soft pencil to avoid cutting through the protective coating of tin (tin-plate) or of zinc (galvanised sheet). Damage to such coatings leads to failure through corrosion.
- Centre punch marks may not control the drill point with sufficient accuracy unless the metal is heavily indented and, even then, total control cannot be guaranteed.
- Centre punching can cause distortion of the work. If the work is thick and the mark is not near the edges, a burr will still be thrown up round the punch mark. When the mark is near the edge of the metal – especially thin metal – the edge of the metal will swell out adjacent to the mark. This can cause inaccuracies if the distorted metal is a datum surface. Thin material, such as sheet metal, may buckle and distort when centre punched. Only the lightest marks should be made in such material.
- The accuracy of a scribed line to rule accuracy is limited to about ±0.5 mm. When using a vernier height gauge this improves to about ±0.1 mm. In practice the accuracy depends upon the condition of the scribing point and the skill of the person using the equipment.

Fig. 5.15 *Checking a casting*

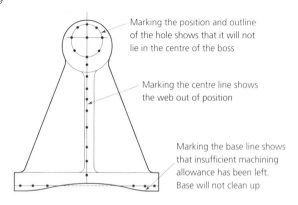

Marking the position and outline of the hole shows that it will not lie in the centre of the boss

Marking the centre line shows the web out of position

Marking the base line shows that insufficient machining allowance has been left. Base will not clean up

5.5 Types of datum

The term datum has already been used several times in this chapter. It has also been described as a point, line or edge from which measurements are taken.

- *Point datum*   This is a single point from which dimensions can be taken when measuring and marking out. For example, the centre point of a pitch circle.
- *Line datum*   This is a single line from which or along which dimensions are taken when measuring and marking out. It is frequently the centre line of a symmetrical component.
- *Edge datum*   This is also known as a *service edge*. It is a physical surface from which dimensions can be taken. This is the most widely used datum for marking out. Usually two edges are prepared at right angles to each other. They are also referred to as *mutually perpendicular* datum edges. These two edges ensure that the distances marked out from them are also at right angles to each other.
- *Surface datum*   For example, this can be the working surface of a surface plate or a marking-out table. As previously described, the surface plate or marking-out table is machined, ground and/or hand scraped to provide a plane surface for marking out. It provides a common datum to support the work and the measuring and marking-out equipment in the same plane. For example, if you set your work with its datum edge on the surface datum of the marking-out table, and you set your surface gauge or scribing block to 25 mm, then the line you scribe on your work will be 25 mm from its datum edge. This is because the datum surface of the foot of your surface gauge and the datum surface of your work are both being supported in the same plane by the surface plate or marking-out table, as shown in Fig. 5.16.

Fig. 5.16   *Marking out from a datum surface*

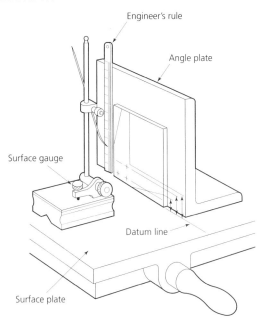

5.5.1 *Coordinates*

The distance from a datum to some feature such as the centre of a hole is called an *ordinate*. In practice, two such dimensions are required to fix the position of a feature on a flat surface. These two ordinates are called *coordinates*. There are two systems of coordinates in common use.

Rectangular coordinates

The feature is positioned by a pair of ordinates (coordinates) lying at right angles to each other and at right angles to the two axes or datum edges from which they are measured. This system requires the preparation of two mutually perpendicular datum edges before marking out can commence. Figure 5.17(a) shows an example of the centre of a hole dimensioned by means of rectangular coordinates. Sometimes rectangular coordinates are called *Cartesian* coordinates after the name of the mathematician who developed this system for fixing points on graphs.

Polar coordinates

In this instance the coordinates consist of a linear (straight line) distance and an angular displacement. Figure 5.17(b) shows how the centre of a hole can be dimensioned using polar coordinates. Dimensioning by this technique is often employed when holes are located around a pitch circle or when machining is taking place on a rotary table. Figure 5.17(c) shows how polar coordinates are used to position holes around a pitch circle. In this example, the linear dimension is the radius of the pitch circle measured from a point datum at its centre. In practice, polar coordinates are rarely used in isolation. They are usually combined with rectangular coordinates, as shown in Fig. 5.17(d).

Fig. 5.17 *Coordinates: (a) rectangular coordinates; (b) polar coordinates; (c) polar coordinates applied to holes on a pitch circle; (d) combined coordinates. (Note: PCD = pitch circle diameter; dimensions in millimetres)*

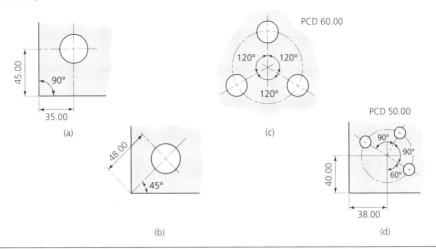

5.6 Techniques for marking out

Having familiarised ourselves with the equipment used for marking out and the types of datum and the systems of coordinates we are likely to meet, it is time to apply all this knowledge to some practical examples.

5.6.1 *Surface preparation*

- Before commencing to mark out a metal surface, the surface must be cleaned and all oil, grease, dirt and loose material removed.
- A dark pencil line shows up clearly on white paper because of the colour contrast. Since scribed lines cut into the metal surface there is very little colour contrast and they do not always show up clearly.
- To make the line more visible, the metal surface is usually coated in a contrasting colour. Large castings are usually whitewashed, but smaller steel and non-ferrous precision components are usually coated with a quick-drying layout 'ink'.
- Avoid using the old-fashioned technique of copper plating the surface of a steel component with a solution of copper sulphate containing a trace of sulphuric acid. Although it leaves a very permanent coating, it can only be used on steels and it is corrosive if it gets on marking-out instruments. The coating can only be removed by using emery cloth or by grinding.
- Layout ink is available in a variety of colours and can be readily applied to a smooth surface using an aerosol can. Shake the can vigorously before use and keep it at the recommended distance from the work. The ink should be applied thinly and evenly. Two thin coats are better than one thick coat. Wait for the ink to dry before marking out. The ink can be removed with a suitable solvent when the component is finished.
- **Safety**   Direct the spray only at the workpiece, never at your workmates. Obey the maker's instructions at all times. Use only if there is adequate ventilation. Avoid breathing in the solvent and the propellant gas.

5.6.2 *Use of a line datum*

Figure 5.18 shows a simple link involving straight lines, arcs and circles. It is symmetrical about its centre line. There are several ways of marking out this component. For the moment, we are going to use the centre line as a *line datum*.

We will assume that we have a flat metal plate of the correct thickness and big enough from which to cut the plate. The following operations refer to Fig. 5.19.

1. Clean the blank (plate) so as to remove all oil, grease and dirt. Remove all sharp corners for safety. Apply a light coating of layout ink to the surface of the blank that is to be marked out. Using a steel rule as a straight edge scribe a centre line along the middle of the plate as shown.
2. Set your dividers to the hole centre distance of 75 mm and step off this distance on the centre line. Leave sufficient room to strike the arcs that form the ends of the links.

Fig. 5.18 *Link. (Dimensions in millimetres)*

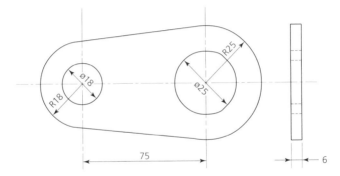

Fig. 5.19 *Marking out from a centre line datum*

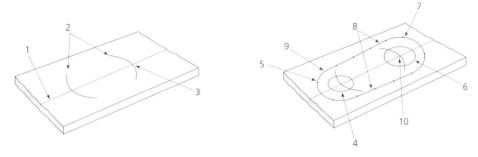

3. Lightly dot punch the intersections of the centre line and the arcs you have struck with your dividers. These centre dots are used to locate the leg of the dividers in the following operations.
4. Set your dividers to 9 mm and scribe in the 18 mm diameter hole.
5. Set your dividers to 18 mm and, using the same centre dot as in 4, strike the smaller end radius.
6. Set your dividers to 12.5 mm and, using the other centre dot, scribe in the 25 mm diameter hole.
7. Set your dividers to 25 mm and, using the same centre dot as in 6, strike the larger end radius.
8. Scribe tangential lines to join the 18 mm and 25 mm end radii using your steel rule as a straight edge to guide the scriber.
9. Preserve the outline by dot punching as described in Section 5.1. The use of witness lines and marks will be discussed further in Section 5.7.
10. Enlarge the hole centre dot punch marks with a centre punch ready for drilling. This completes the marking out.

5.6.3 *Use of a single edge datum*

The following sequence of operations refers to Fig. 5.20. It assumes that the metal blank from which we are going to make the link has at least one straight edge. This would be the case if the blank was sawn from a piece of 75 mm by 6 mm bright drawn, low-carbon steel.

1. Clean the sawn blank so as to remove all oil, grease and dirt. Remove all sharp corners for safety. Apply a light coat of layout ink. Use a steel rule as a straight edge to check the selected datum edge for flatness and straightness. Carefully remove any bruises with a fine file.
2. Scribe the centre line parallel to the datum edge using odd-leg calipers as shown.
3. Scribe the first centre line at right angles to the datum edge using a try-square to guide the scriber point, leaving room for striking the arc that forms the end of the link.
4. Measure and mark off the centre distance to the second hole either by using your rule and scriber as shown or by stepping off the distance with dividers set to 75 mm as in the previous example.
5. Scribe the second hole centre line at right angles to the datum edge using a try-square. Dot punch the centre points.
6. The remaining operations are the same as 4 to 10 inclusive in the previous example.

Fig. 5.20 *Marking out from a datum edge*

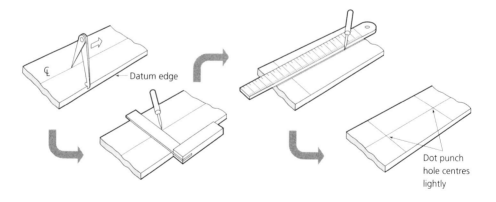

A variation on the theme is shown in Fig. 5.21. By clamping the blank to an angle plate we get the same effect as having a pair of mutually perpendicular datum edges. This enables us to scribe the centre lines at right angles to each other without the use of a try-square. By using a vernier height gauge in place of the scribing block shown, the centre distance can be marked out much more accurately than by using a rule, as in the previous example. The plate can be clamped by using small G-clamps as shown or by using toolmaker's clamps.

Fig. 5.21 *Use of angle plate (a) to provide mutually perpendicular datum surfaces; (b) toolmaker's clamp*

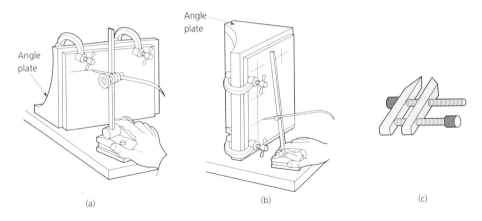

(a)   (b)   (c)

5.6.4 *Mutually perpendicular datum edges*

This time we will assume that our blank has two datum edges that are at right angles to each other, i.e. they are mutually perpendicular. The general set up for marking out is shown in Fig. 5.22 and the following sequence of operations refers to Fig. 5.23.

Fig. 5.22 *Marking out from a datum surface – the surface plate provides the datum surface; all measurements are made from this surface; all lines scribed by the scribing block will be parallel to this surface*

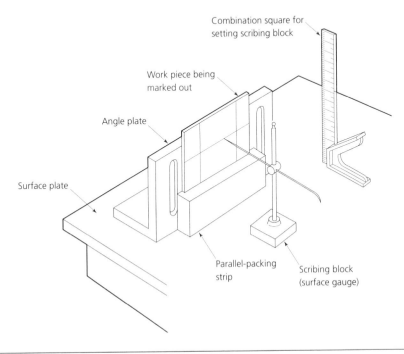

1. File or machine two edges at right angles to each other and to the surface being marked out. Remove all sharp edges, oil, grease and dirt from the blank and apply a light coat of layout ink.
2. The blank is placed on its end datum edge on a marking-out table as shown. A precision ground, parallel packing block is used to raise the work to a convenient height. The thickness of the packing must be measured and allowed for when setting the scribing point. The point of the scriber is set to the combination rule. Make sure the datum end of the rule is in contact with the surface of the marking-out table. A line is now scribed on the blank at this setting. The scribing point is then raised by 75 mm and a second line is scribed as shown.
3. The blank is then turned through 90° so that it rests on the other datum edge. This enables the remaining centre line to be scribed at right angles to the first two. The hole centres are where the lines intersect. Dot punch these centres lightly. The marking out of the link is completed as described in operations 4 to 10 inclusive in the first example. If greater accuracy is required a vernier height gauge is used in place of the scribing block. This was shown in a previous example.

Fig. 5.23 *Marking out procedure when using a datum surface*

File or machine up two edges at right angles (perpendicular) to each other and at right angles to the face being marked out

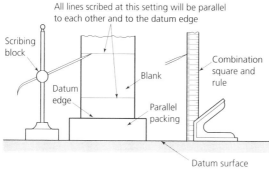

Blank is placed on datum edge on surface plate (datum surface). In this example parallel packing is used to raise the blank to a convenient height. The scribing block is set to a combination square and rule. The setting is transferred to the blank. The line so scribed will be parallel to the datum surface and therefore parallel to datum edge of the blank

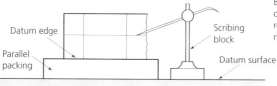

Blank is turned through 90° so that it rests on the other datum edge. This enables the remaining centre line to be scribed in at right angles to the first two

5.6.5 *Use of a point datum and tabulated data and dimensions*

Figure 5.24 shows a component that has been drawn using rectangular coordinates and absolute dimensioning for the hole centres. Each hole centre then becomes a *point datum* for the clusters of small holes. To avoid confusion, the large number of repeated dimensions for the holes have been tabulated. This is referred to as tabulated data.

Fig. 5.24 *Tabulated data*

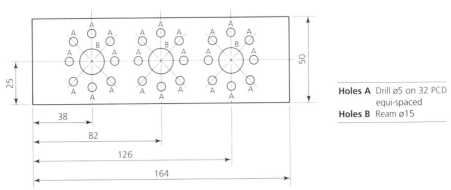

Holes A Drill ø5 on 32 PCD
equi-spaced
Holes B Ream ø15

Because the major hole centres have been dimensioned using rectangular coordinates, they can be marked out as described previously. Either the blank can be clamped to an angle plate which is rotated at right angles in order to scribe lines at right angles, or the blank can be prepared with a pair of mutually perpendicular datum edges. This enables the blank to be turned through 90° without having to turn the angle plate on end.

The dot punch marks at the intersection of the centre lines are used to locate the dividers. These are used to mark out the outline of the 15 mm diameter holes and also the 32 mm diameter pitch circles for the smaller holes. Since there are eight equispaced holes in each cluster, they will be at 45° to each other. We can mark out their centre positions using the square and mitre head from the combination set, as shown in Fig. 5.25(a). Had their been six holes, their chordal distance would have been the same as the pitch circle radius. Therefore, after marking out the pitch circles, the hole centres could have been stepped off with the dividers at the same setting, as shown in Fig. 5.25(b).

Sometimes a drawing has to satisfy a family of similar components that only vary in size but not in shape. Such an example is shown in Fig. 5.26. This drawing has tabulated dimensions for the overall length and the hole centres. The width and thickness of the component remains constant and the holes are located on the centre line, which is also constant.

Fig. 5.25 *Marking out holes on a pitch circle: (a) use of combination square; (b) use of dividers*

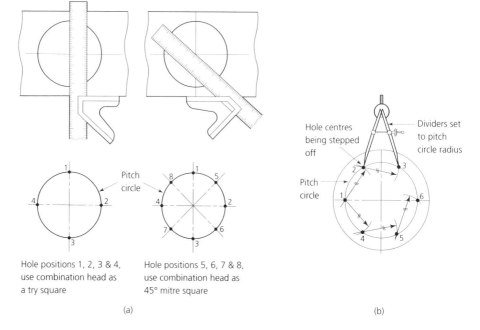

Hole positions 1, 2, 3 & 4, use combination head as a try square

Hole positions 5, 6, 7 & 8, use combination head as 45° mitre square

(a)

(b)

Fig. 5.26 *Tabulated dimensions*

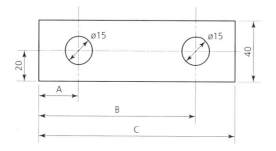

Component	A	B	C
1/316/A	25	100	125
2/316/A	30	120	150
3/316/A	50	200	250

5.7 Minimising inaccuracies

Inaccuracies can occur in two ways when:

- The lines themselves are not correctly positioned.
- There are difficulties in working to the scribed lines.

5.7.1 *Use of appropriate equipment*

A range of marking-out equipment has been introduced in this chapter together with a corresponding range of applications. It is important to match the equipment used to the job in hand if inaccuracies are to be minimised. For example:

- A scriber and steel rule should only be used if the accuracy required is coarser than ±0.5 mm. This is adequate if the scribed line is only for guidance and the finished size is to be controlled by the micrometer dials on a machine or by the use of precision-measuring equipment, as previously described in Chapter 4.
- A vernier height gauge should only be used when greater accuracy is required (e.g. toolmaking). It would be a waste of time and expensive equipment to use this instrument if a scriber and rule would suffice. A vernier height gauge can be read to an accuracy of ±0.01 mm but, because of the width of the scribed line, the working accuracy is usually taken as ±0.1 mm. In any case, when working to a scribed line it would be difficult to achieve this sort of accuracy.
- A plain protractor, such as the protractor head of a combination set, is suitable for angles whose accuracy is coarser than ±0.5°.
- A vernier protractor (see Chapter 4) should be used for finer angular measurements. This instrument has a reading accuracy of 5 minutes of arc.

5.7.2 *Condition and care of equipment*

Marking-out equipment should be kept in good condition if inaccuracies are to be avoided.

- As has been mentioned previously, the points of scribers and dividers should be kept needle sharp by regular dressing with a fine oil slip. This is shown in Fig. 5.27(a). Do not sharpen by grinding; the heat generated will soften the scribing point.

Fig. 5.27 *Care of marking out equipment: (a) sharpening scribing points; (b) sharpening height gauge scribing blades; (c) sharpening centre and dot punches; (d) correct dot and central punch point configurations*

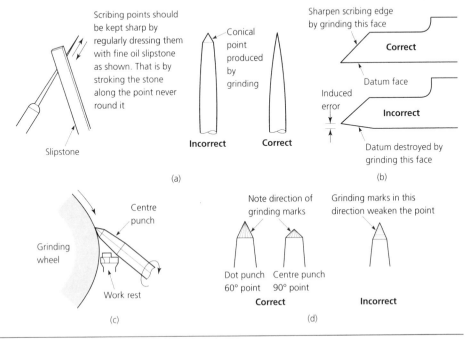

- The nib of a vernier height gauge should be sharpened by carefully grinding, as shown in Fig. 5.27(b). Use a free cutting wheel to avoid overheating and softening the scribing edge. Use an appropriate silicon carbide (green grit) wheel if the nib is tungsten carbide tipped.
- When sharpening the point of a dot punch or a centre punch, the punch is presented to the abrasive wheel, as shown in Fig. 5.27(c). This ensures that the grinding marks run down the point and not round it, as shown in Fig. 5.27(d). The techniques that will be shown in Fig. 5.29(c) gives the point greater strength and a longer life.
- Rules should be kept clean. The datum end should be protected and *never* used as a makeshift screwdriver or for shovelling swarf out of the T-slot of a machine tool bed. The edges of a rule must also be kept in good condition if it is to enable straight lines to be scribed or it is to be used as a straight edge.
- Try-squares must also be treated carefully and cleaned and boxed when not in use. They should never be dropped, mixed with other tools or used for any purpose other than that for which they are designed. They should be tested for accuracy from time to time. A simple test was shown in Fig. 4.36.
- Angle plates must be kept clean and free from bruises. Bruises not only prevent proper contact between the angle plate and the work it is supporting, they also cause damage to the surface of the marking-out table on which they are supported.
- Surface plates and marking-out tables must also be treated with care as they provide the datum from which other dimensions are taken. Heavy equipment must be slid onto the table from the edge, and lighter equipment must be placed onto it carefully. If a surface plate or a marking-out table becomes scratched or damaged in any way, the damaged area must be carefully stoned or scraped flat again. Always cover the table when it is not in use.
- Vee-blocks must be kept boxed in pairs as originally supplied. They are made in matched pairs and must be kept together for the whole of their working lives. Vee-blocks from two different sets will not necessarily support a shaft parallel to the datum surface on which the blocks are supported.
- The above notes refer not only to the items of equipment specifically mentioned but also to all similar marking equipment and to marking-out and measuring equipment in general. Table 5.1 summarises the more usual causes of faults and inaccuracies when marking out.

5.7.3 Cutting and limit lines

These are lines that indicate where the workpiece is to be cut when sawing and shearing, or the limit to which the metal is to be cut away when filing or machining. Even when these lines, produced during marking out, are correctly and accurately scribed it is still difficult to work to them accurately.

The concept of dot punching scribed lines to preserve them has already been introduced. Let us now examine this technique more closely. Scribed lines are often marked with a dot punch, as shown in Fig. 5.28. Small dot punch or 'pop' marks are made along a straight line at about 20 to 25 mm intervals and at corners, as shown in Fig. 5.28(a). They should be closer together around curves and complex profiles. Be careful to locate the point of the dot punch accurately on the scribed line when dot punching. If the scribed line should become defaced or erased, it can be restored by using a scriber to connect the dot marks again.

Table 5.1 *Faults and inaccuracies when marking out*

Fault	Possible cause	To correct
Inaccurate measurement	Wrong instrument for tolerance required	Check instrument is suitable for tolerance required
	Incorrect use of instrument	Improve your technique
Scribed lines out of position	Parallax (sighting) error	Use the scriber correctly
	Rule not square with datum edge	Use a datum block (abutment)
Lines not clear	Scribing point blunt	Sharpen the point of the scriber
	Work surface too hard	Use a surface coating (spray-on lacquer)
	Scribing tool lacks rigidity	Use only good-quality tools in good condition
Corrosion along scribed lines	Protective coating (tin plate) cut by using too sharp a scribing point	Use a pencil when marking coated materials
Component tears or cracks along scribed line when bent	Scribed line and direction of bend parallel to grain of material	Bend at right angles to grain of the material
	Scribed line cut too deeply	Mark bend lines with a pencil
Circles and arcs irregular and not clear	Scribing points blunt	Resharpen
	Instruments not rigid	Use only good quality dividers or trammels of correct size for job
	Centre point slipping	Use a dot punch to make a centre location
Centre punch marks out of position	Incorrect use of punch	Position punch so that point is visible and then move upright when point is correctly positioned
	Scribed lines not sufficiently deep to provide a positive point location	Ensure point can click into the junction of the scribed lines

Fig. 5.28 *Preserving a scribed line: (a) dot punching; (b) witness marks*

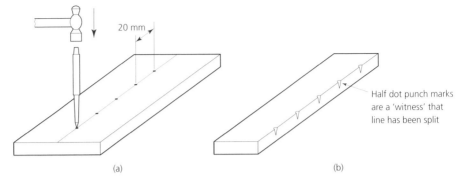

20 mm

Half dot punch marks
are a 'witness' that
line has been split

(a)   (b)

Another use for these 'pop' marks is as an aid to machining. If you machine down to a scribed line so as to 'split the line', there will be no line left to prove that you have worked accurately to the line. However, if the line has dot marks along it, and you have accurately split the line, then half the marks will still be visible to prove the accuracy of your work, as shown in Fig. 5.28(b). For this reason such marks are often called *witness marks*.

5.7.4 *Round holes – size and position*

In theory, all you need when marking out hole centres ready for drilling is a centre punch mark at the intersection of the centre lines as a guide for the drill point. Unfortunately drills have a habit of 'wandering', especially when starting a large drill with a centre punch mark. Therefore, it is usual to also mark out the circle representing the hole, as shown in Fig. 5.29(a), using dividers. The hole is then dot punched. If the hole is drilled accurately, the dot punch marks should be split. However, this assumes that:

- The centre lines are accurately marked out.
- The centre punch mark is exactly at the intersection of the centre lines.
- The dividers are exactly set to the hole radius.
- The dividers do not 'wander' in the centre punch mark.
- The 'pop' marks around the circle are accurately positioned.

Fig. 5.29 *Marking out round holes*

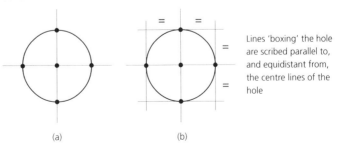

Lines 'boxing' the hole
are scribed parallel to,
and equidistant from,
the centre lines of the
hole

(a)   (b)

These are a lot of assumptions. For this reason it is better to 'box' the hole, as shown in Fig. 5.29(b). Whilst the hole centres are being accurately marked out using rectangular coordinates, the vernier height gauge can also be used to accurately scribe lines either side of the centre lines at a distance equal to the radius of the hole. This produces an accurate box within which the drilled hole should lie.

5.7.5 *Guide lines*

Guide lines and witness lines are also used in conjunction with straight cutting lines, as shown in Fig. 5.30. The guide line is scribed parallel to the cutting line or the limit line and it is positioned on the waste material side, as shown in Fig. 5.30(a), therefore the guide line will be removed during machining. In the case of a drilled hole or a bore, the guide line is a circle slightly smaller than the finished size of the hole or bore. This is shown in Fig. 5.30(b).

The reason for guide lines is to provide a visual check that the work is correctly set and that machining is being carried out parallel to the cutting line or the limit line. This enables adjustments and corrections to be made before cutting to the final size. For this reason more than one guide line may be provided.

Fig. 5.30 *Guide lines*

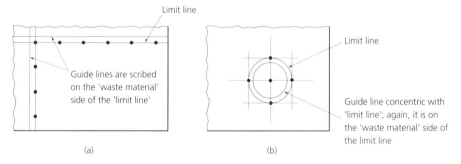

5.7.6 *Witness lines*

Witness lines are scribed parallel to the cutting or the limit line on the opposite side to the guide line, as shown in Fig. 5.31. Therefore, when cutting is complete, the lines should still be present. They are used in conjunction with or in place of the dot-punch witness marks described earlier. If cutting or machining has been performed successfully and correctly, the witness line should be parallel to the edge of the component and the correct distance from it. It remains as a witness to the accuracy of the fitting or machining processes used.

Figure 5.32(a) shows how witness lines are applied to a round hole. The circular or boxed lines can act as a *witness* to the size and position of the drilled hole, but what happens if the hole should wander? It is too late once it has been drilled. The answer is to also provide a *guide line* by marking out a circle that is slightly smaller than the finished

Fig. 5.31 *Witness lines*

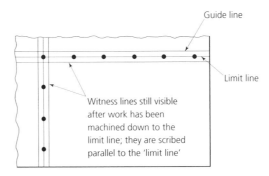

Guide line

Limit line

Witness lines still visible
after work has been
machined down to the
limit line; they are scribed
parallel to the 'limit line'

Fig. 5.32 *Witness lines for round holes: (a) guide lines; (b) hole correct size and centred accurately – it just touches the witness lines; (b) hole centred accurately but oversize – it cuts all the witness lines equally; (d) hole off centre and cuts two witness lines – it would still be off centre if it cut only one witness line*

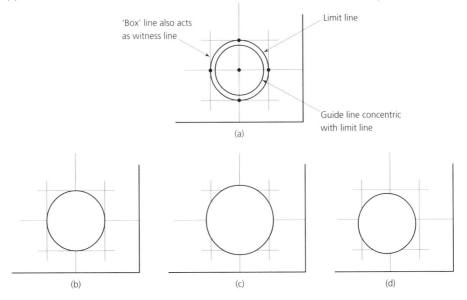

'Box' line also acts
as witness line

Limit line

Guide line concentric
with limit line

(a)

(b)   (c)   (d)

diameter but slightly larger than the diameter of the pilot drill hole. Figure 5.32(b) shows a correctly drilled hole, whilst Fig. 5.32(c) and Fig. 5.32(d) show how the witness lines expose incorrectly positioned and/or sized holes. If the hole drilled by the pilot drill is central in the *guide line* circle, then the drill used for sizing the hole should also be central if it has been correctly ground.

In the pilot drill is not correctly centred in the guide line (circle), then you have no alternative but to bore out the hole. There is no drilling or reaming process that can correct the position of a hole. Fortunately drilling is recognised as being a quick rather than an accurate means of producing holes. Therefore drill holes are usually designed to be slightly oversize to allow for any slight wander. The positioning and production of precision holes will be considered in *Fundamentals of Mechanical Engineering*.

5.7.7 *Line enhancement*

The use of layout inks has already been discussed. However, their use can be messy and they tend to dissolve away with some coolants during cutting or be scratched away by the swarf during machining operations. For this reason it is sometimes better to enhance the line itself to give a colour contrast. One of the following substances can be rubbed into the scribed line, as shown in Fig. 5.33.

- Engineer's blue will enhance the clarity of scribed lines on bright shiny metals.
- Chalk powder will enhance the clarity of scribed lines on dull dark metals, such as grey cast iron.
- Graphite from crushed pencil leads can be used with good effect to enhance lines scribed on non-metallic materials.

Fig. 5.33 *Line enhancement*

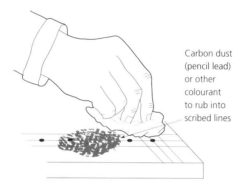

Carbon dust
(pencil lead)
or other
colourant
to rub into
scribed lines

5.8 Storage of marking-out equipment

The use and care of marking-out equipment has already been introduced. Here are some further hints and tips concerning the care and storage of marking-out equipment.

- Sharp-pointed instruments should have their points guarded for safety. They can cause some nasty cuts that can easily become infected in a workshop environment. Push the sharp points into a suitably sized cork when the instrument is not in use. This protects the point and you.
- Marking-out and measuring equipment must never be stored with cutting tools. Don't store measuring and marking-out equipment in a drawer along with hammers, chisels and files. The datum edges of the measuring and marking-out equipment will soon be damaged and the equipment will quickly become inaccurate.

- Marking-out and measuring instruments should be kept in instrument cupboards to protect them from the general run of equipment found in a workshop stores.
- Clean all marking-out equipment after use. The sweat from your fingers will quickly corrode the instruments. Wipe with a good-quality light machine oil to protect against atmospheric moisture if the equipment is not going to be used for some time.
- Expensive instruments, such as vernier height gauges, are usually supplied in fitted, polished wood cases together with their accessories. Keep them in their boxes when not in use. These boxes are fitted with support cradles to prevent the instrument beams from becoming distorted.
- All equipment should be checked when it is returned to the stores and sent for maintenance, if required, before being re-issued.
- Your personal marking-out equipment should be kept in a proprietary tool cabinet. Marking-out and measuring equipment should be kept in separate drawers away from cutting tools and heavier equipment such as hammers and clamps.

5.9 Reporting damaged and faulty equipment

You cannot be expected to carry out good-quality work with defective equipment. Neither can any one else.

- If you accidentally damage a piece of equipment or it becomes defective whilst you are using it, return it to the stores and report its condition. It's unfair on your workmates for it to be re-issued without maintenance and repair.
- Some firms have a reporting system for logging equipment faults. Use it.
- If you are issued with damaged or faulty equipment, return it to the stores and ask for it to be changed. If you can get no satisfaction from the stores staff, seek advice from your instructor or your supervisor. No matter how frustrating, do not become aggressive this will only aggravate the situation and you may have to rely on the good-will of the stores staff at another time.
- Always complete the documentation when booking equipment into and out of the stores. This prevents equipment becoming mislaid leading to confusion, and frayed tempers later.

EXERCISES

5.1 *Marking out and marking-out equipment*
  (a) (i) List the reasons for marking out components ready for manufacture.
      (ii) Explain why mass produced components are not marked out prior to manufacture.
  (b) List the advantages and limitations of manual marking out in terms of accuracy and possible damage to the surfaces of the workpiece.
  (c) Complete Table 5.2. It has been started to give you a guide.

Table 5.2 *Exercise 5.1(c)*

Technique	Equipment required
Straight lines	Rule and scriber
Circles and arcs	
Lines parallel to an edge (not using a surface plate)	
Lines parallel to a surface plate	
Lines parallel to angle sections	
Lines along shafts parallel to each other and to the axis of the shaft	
Lines perpendicular to an edge	

5.2 *Techniques for marking out*

(a) Draw up an operation schedule for marking out the component shown in Fig. 5.34, using its centre line as a datum. List the marking-out operations in the correct order and the equipment used.

Fig. 5.34 *Exercise 5.2(a). (Dimensions in millimetres)*

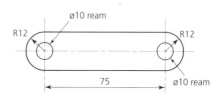

(b) With the aid of sketches, explain how lines can be scribed parallel and perpendicular to a surface plate datum.

(c) With the aid of sketches, show how lines can be drawn parallel to angle sections.

(d) State the purpose of a box square and, with the aid of sketches, show how it can be used.

(e) Describe the difference between a dot punch and a centre punch and explain their uses.

5.3 *Types of datum*

(a) Copy and complete Figure 5.35.

Fig. 5.35 *Exercise 5.3(a). (Dimensions in millimetres)*

Type of datum	Sketch of example
Edge of datum	
Line of datum	
	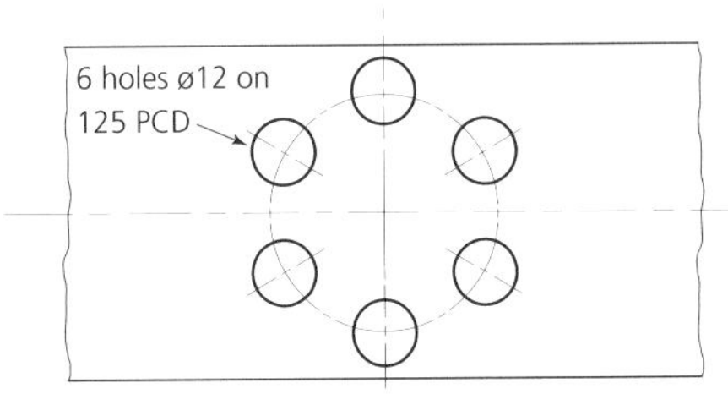 6 holes ø12 on 125 PCD

(b) With the aid of sketches, describe how the component shown in Fig. 5.36 can be marked out on a surface table using the edge marked AA as a datum.

(c) With the aid of sketches describe how a scribed line can be 'protected' using dot-punch marks. Also describe how these dot-punch marks can act as a 'witness' that a fitter or a machinist has worked correctly to a scribed line.

5.4 *Minimising inaccuracies when marking out*

(a) Explain what is meant by *parallax errors* when marking out using a steel rule. How can such errors be minimised?

(b) Describe two ways in which a scribed line can be made to show up more clearly.

(c) Describe **two** ways (other than those in (a) and (b)) by which marking-out inaccuracies can be minimised.

5.5 *Care of marking-out tools and equipment*

(a) With the aid of sketches, describe how the scribing points/edges of the following marking-out tools should be sharpened:

    (i) divider points

    (ii) vernier height gauge nib

    (iii) dot-punch point

(b) Describe how marking-out tools and measuring instruments should be cared for and stored in order to maintain their accuracy and keep them in good condition.

Fig. 5.36 *Exercise 5.3(b). (Dimensions in millimetres)*

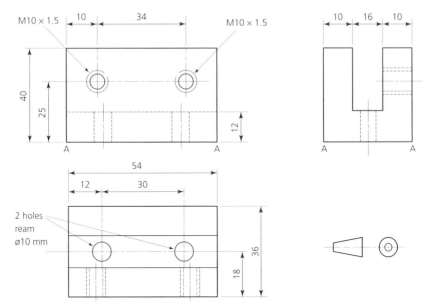

6 Engineering materials

When you have read this chapter, you should understand:

- How to define the basic properties of engineering materials.
- How to identify correctly and select a range of engineering metals and alloys.
- How to identify correctly and select a range of non-metallic materials suitable for engineering applications.
- How to check for the common defects that occur in engineering materials by 'non-destructive testing'.
- The principles of 'destructive' testing to establish the mechanical properties for a given material.

6.1 Elements, compounds and mixtures

There are 93 known elements which occur naturally. They make up everything in the world, both living and non-living. So that we can understand our materials more readily, let's first remind ourselves of some basics.

6.1.1 *Elements*

These are substances that cannot be broken down into simpler substances by chemical or electrochemical reactions. Water is not an element because you can break it down into the gases hydrogen and oxygen from which water is made. However, hydrogen and oxygen cannot be broken down, they are chemical elements.

6.1.2 *Compounds*

When elements combine together to form compounds:

- Heat is either taken in or given out
- The compound is different in appearance and properties to the elements from which it is made.

For example, table salt (chemical name: sodium chloride) is a chemical compound produced when the silvery and highly reactive metal sodium is burnt in the poisonous gas chlorine. Both sodium and chlorine are elements. Sodium chloride is a white crystalline or powdery substance and is a compound. Water (a liquid) is formed when hydrogen gas burns in oxygen (both gases). Water is a compound, hydrogen and oxygen are elements.

6.1.3 *Mixtures*

When you mix elements together, or elements and compounds together, or just compounds together, without a chemical reaction no new substances are made, no heat is taken in or given out, no change in appearance or new properties are apparent. For example, a mixture of iron filings and sand will remain just a mixture of iron filings and sand. The iron filings can be removed with a magnet to separate the mixture into its component materials. No chemical processes are required.

6.1.4 *States of matter*

Almost all matter can exist in three physical states by changing its temperature in appropriate conditions. These states are solid, liquid and gas.

- Ice is *solid* water and exists below 0 °C.
- Water is a *liquid* above 0 °C and below 100 °C.
- Steam is water *vapour* above 100 °C and becomes a *gas* as its temperature is raised further (superheated).

Metals such as brass, copper or steel are solid (frozen) at room temperatures but become liquid (molten) if heated to a sufficiently high temperature. If they are heated to a high enough temperature they will turn into a gas. On cooling, they will first turn back to a liquid and then back to a solid at room temperature. Providing no chemical change takes place (oxidation of the metal through contact with air at high temperatures) we can change substances backwards and forwards through the three states by heating and cooling as often as we like.

There are exceptions; for example, when a thermosetting plastic has been heated it undergoes a chemical change called 'curing'. Once 'cured' it can never again be softened or turned into a liquid by heating. It can be destroyed by overheating. Another example is the non-metallic element iodine. When heated this *sublimes* directly from a solid to a vapour without becoming a liquid.

With the exception of mercury, all metals are solid at normal working temperatures. Metals can be melted into liquids by heating them sufficiently. In the liquid state they can be cast to shape in moulds. For convenience, the fuels used in engineering for welding or for heating furnaces are liquids (oil fuels) or gases such as propane, acetylene, methane (natural gas), etc.

6.2 Properties of materials

So that we can compare and identify engineering materials, it is important to understand the meaning of their more common properties. For example, we cannot say that one metal is harder than another unless we know what is meant by the term 'hardness'.

6.2.1 *Strength properties*

Tensile strength
This is the ability of a material to withstand a stretching load without breaking, and is shown in Fig. 6.1(a). The load is trying to stretch the rod, therefore the rod is said to be in a state of tension. It is being subjected to a tensile load. To resist this load without breaking, the material from which the rod is made needs to have sufficient *tensile strength*.

Compressive strength
This is the ability of a material to withstand a squeezing load without breaking, and is shown in Fig. 6.1(b). The load is trying to squash (crush or compress) the material from which the component is made; therefore the component is said to be in a state of compression. It is being subjected to a compressive load. To resist this load without breaking, the material from which it is made needs to have sufficient *compressive strength*.

Fig. 6.1 *Material properties – strength: (a) tensile strength; (b) compression strength; (c) shear strength; (d) impact strength*

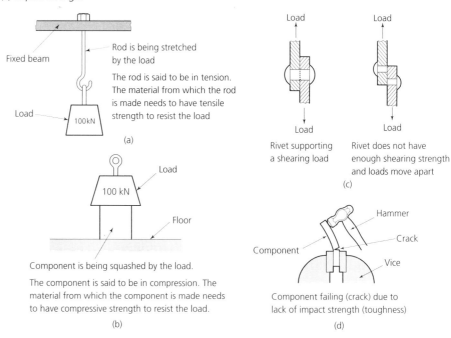

Shear strength

This is the ability of a material to withstand an offset load without breaking (shearing), and is shown in Fig. 6.1(c). The loads are trying to pull the joint apart and the rivet is trying to resist them. The loads are not in line, but are offset. The rivet is subjected to a shear load. The material from which it is made must have sufficient *shear strength* or the rivet will fail as shown, and the loads will move apart. The rivet is said to have *sheared*. The same effect would have occurred if the loads had been pushing instead of pulling. Note: riveted joints should always be designed so that the load acts in shear across the shank of the rivet as shown. It must never pull on the heads of the rivet. The heads are only intended to keep the rivet in place.

Toughness

This is the ability of a material to withstand an impact or 'hammering' load, and is shown in Fig. 6.1(d). The constant hammering (impact loading) is causing the metal to crack. To resist this impact loading without breaking, the material from which it is made needs to have sufficient *toughness*. Strength and toughness must not be confused. Strength refers to tensile strength – the ability to withstand an axial pulling load. For example, when you buy a rod of high-carbon steel (e.g. silver steel) it is in the soft condition and it is strong and tough. It has a relatively high tensile strength and it will withstand quite a lot of hammering (impact loading) before it cracks.

- If you heat this metal to 'cherry-red' and cool it quickly in cold water (quench it), it will become very hard.
- Its *tensile strength* will also have greatly *increased*.
- It will also have become very *brittle*.
- It will now break with only a light tap with a hammer – it can *no longer resist impact loads* – it has *lost its toughness*.

Brittleness

Brittleness is the opposite of toughness. It is the ability to shatter when subjected to impact – as a glass window behaves when struck by a cricket ball.

6.2.2 *Miscellaneous properties*

Elasticity

This property enables a material to change shape under load and to return to its original size and shape when the load is removed. Components such as springs are made from elastic materials, as shown in Fig. 6.2(a). Note that springs will only return to their original length if they are not overloaded.

Plasticity

This property enables a material to deform under load and to retain its new shape when the load is removed. This is shown in Fig. 6.2(b). The coin is made from a copper alloy that is relatively soft and plastic. It takes the impression of the dies when compressed between them, and retains that impression when the dies are opened. When the deforming force is tensile, the property of plasticity is given the special name *ductility*. When the deforming force is compressive, the property of plasticity is given the special name *malleability*.

Ductility

As stated above, this property enables a material to deform in a plastic manner when subjected to a tensile (stretching) force; for example, when wire drawing as shown in Fig. 6.2(c). To combine *tension* and *plasticity* results in the need for a *ductile* material.

Malleability

As stated above, this property enables a material to deform in a plastic manner when subjected to a compressive (squeezing) force; for example, when forging or as when rivet heading, as shown in Fig. 6.2(d). The material must retain its shape when the compressive force is removed, so it must be plastic. To combine *compression* and *plasticity* results in the need for a *malleable* material.

Fig. 6.2 *Material properties – flow: (a) elasticity; (b) plasticity; (c) ductility; (d) malleability*

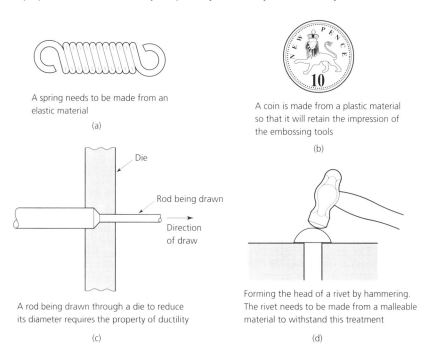

A spring needs to be made from an elastic material

(a)

A coin is made from a plastic material so that it will retain the impression of the embossing tools

(b)

Die

Rod being drawn

Direction of draw

A rod being drawn through a die to reduce its diameter requires the property of ductility

(c)

Forming the head of a rivet by hammering. The rivet needs to be made from a malleable material to withstand this treatment

(d)

Rigidity

This property is also referred to as *stiffness*. This is the ability of a metal to retain its original shape under load – that is, to resist plastic or elastic deformation. Cast iron is an example of a rigid material. Because it is rigid and because it can be cast into intricate shapes, it is a good material to use when making the beds and columns for machine tools.

Hardness

This is the ability of a material to resist scratching and indentation. Figure 6.3(a) shows a hard steel ball being pressed into the surface of two pieces of material using the same standard load. When pressed into a hard material, the ball only makes a shallow indentation. When pressed into a soft material, under the same test conditions, the ball sinks into the material further and makes a deeper impression. This is the basic principle of all standard hardness tests.

Corrosion resistance

This is the ability of a material to withstand chemical or electrochemical attack. A combination of such everyday things as air and water will chemically attack plain carbon steels and form a layer of rust on its exposed surfaces. Stainless steels are alloys of iron together with carbon, nickel and chromium. They will resist corrosion (see Fig. 6.3(b)). Many non-ferrous metals are corrosion resistant, which is why we use copper for water pipes, and zinc or lead for roofing sheets and flashings.

Electrical conductivity

This is the ability of a material to conduct electricity. The better it conducts electricity the higher its *conductivity* and the lower its *resistivity*. Copper is the most widely used material for electrical conductors because:

- It is readily available.
- Its conductivity is second only to silver.
- It is corrosion resistant.
- It can easily be soldered to cable lugs.
- It has a relatively high mechanical strength.
- It is highly ductile and can easily be drawn out into the finest of wires.
- It is flexible.

Other metals are also used for conductors. For example, aluminium is used for conductors, tungsten is used for electric light bulb filaments, nickel-chromium alloys are used for heating elements, and manganin alloy is used for resistance elements in electrical instruments. Carbon is the only non-metallic conductor and is used for the 'brushes' for electrical motors and generators. It is also used for the small resistors found in electronic equipment. It has a low conductivity (high resistance) compared with metals and metal alloys.

Materials with a very poor conductivity (very high resistance – much higher than carbon) are used as insulating materials. Plastics and rubber are widely used as insulating materials. Figure 6.3(c) shows a typical electric cable. Remember: there is no such thing as a perfect conductor or a perfect insulator, and the performance of conductors and insulators deteriorates with any rise in temperature.

Heat conductivity

This is the ability of a material to conduct heat. As for electricity, metals are good conductors of heat and non-metals are poor conductors of heat. Figure 6.3(d) shows an electrically heated soldering iron. The bit is made of copper because this is the best of the common metals for conducting heat. It conducts the heat from the heating element to the joint to be soldered. Copper also has an affinity for soft solder so it can easily be 'tinned'. The handle is made of wood or plastic as these materials are easily shaped and are poor conductors of heat. They are heat-insulating materials. They keep cool and are pleasant to handle.

Refractoriness

You must not mix up poor heat conductivity with good refractory properties. *Refractory* materials are largely unaffected by heat. The fire bricks used in furnaces are refractory materials. They do not burn or melt at the operating temperature of the furnace. They are also good heat-insulating materials. However, the plastic or wooden handle of the soldering iron shown in Fig. 6.3(d) had good heat-insulating properties but these are not refractory materials since they are destroyed by high temperatures.

Fig. 6.3 *Material properties – miscellaneous: (a) hardness; (b) corrosion resistance; (c) conductivity (electrical); (d) conductivity (heat); (e) fusibility*

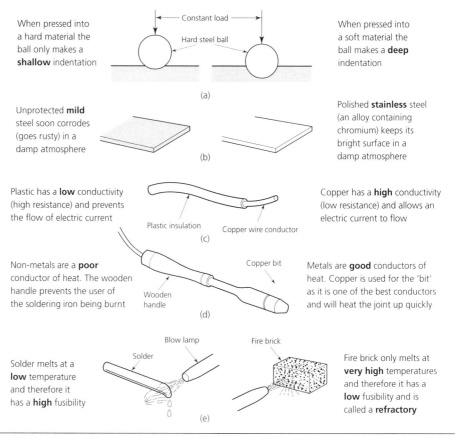

Fusibility

This is the ease with which materials melt. Soft solders melt at relatively low temperatures, other materials melt at much higher temperatures. Figure 6.3(e) shows the effect of turning a gas blowpipe onto a stick of soft solder. The solder quickly melts. The same flame turned onto a block of steel makes the steel hot (possibly red hot) but does not melt it.

- The soft solder *melts at a low temperature* because it has *high fusibility*.
- The steel will glow red hot but *will not melt* in the flame of the blowpipe because it has *low fusibility* and will only melt at a very *high temperature*.

6.2.3 *Magnetic properties*

Magnetic materials are mainly based on the metal iron. Therefore, they are referred to as *ferro-magnetic* materials because iron is the basis of all *ferrous* metals (see Section 6.4). The only other metals that are strongly magnetic are cobalt, gadolinium and nickel.

If an electric current is made to flow through a coil of wire (a solenoid), as shown in Fig. 6.4(a), a magnetic field is produced. If a piece of low-carbon steel is inserted into the magnetic field of the coil, as shown in Fig. 6.4(b), it will become magnetised. When

Fig. 6.4 *Material properties – magnetic: (a) magnetic field of a solenoid; (b) magnetic field around an electromagnet; (c) permanent magnets*

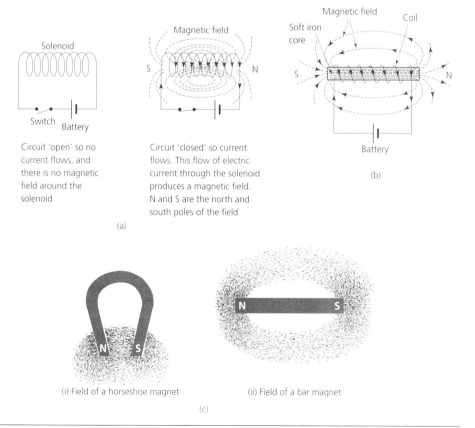

Circuit 'open' so no current flows, and there is no magnetic field around the solenoid

Circuit 'closed' so current flows. This flow of electric current through the solenoid produces a magnetic field. N and S are the north and south poles of the field

(a)

(b)

(i) Field of a horseshoe magnet

(ii) Field of a bar magnet

(c)

removed from this magnetic field it will lose most of its magnetism. If the current is turned off so that the magnetising field is destroyed, or the piece of steel is removed from the magnetic field, it immediately loses almost all of its magnetism. Therefore it is said to be a *'soft' magnetic material*. Materials that can be magnetised or demagnetised in this way are used in powerful electromagnets that are suspended from cranes for lifting scrap iron and steel. Electromagnets are also used in relays, and industrial electric motor starter switches.

If a piece of quench-hardened high-carbon steel is now placed in the magnetic field of the solenoid, it will also become magnetised. If the current is switched off or the piece of steel is removed from the solenoid, it will remain magnetised. It has become a *permanent magnet*. Permanent magnets are made from *'hard' magnetic materials*. Very powerful permanent magnets are made from alloys containing iron together with one or more of the following: cobalt, chromium, tungsten, aluminium, copper, nickel, niobium and titanium. Permanent magnet chucks are used for work holding on grinding machines.

A typical *bar magnet* and a typical *horseshoe magnet* are shown in Fig. 6.4(c). These are both examples of permanent magnets.

Remember that magnets will only attract other magnetic materials. For example, you can only hold a steel or iron component on a magnetic chuck. A copper, brass, aluminium or other non-magnetic material would not be attracted to the magnet and cannot be held on a magnetic chuck.

6.2.4 *Heat treatment*

Heat treatment is dealt with in detail in Chapter 7. However, since the purpose of all heat-treatment processes is to alter the properties of engineering materials, it is necessary to refer to some basic heat-treatment terms, from time to time, in relation to the properties of materials in this chapter.

Heat treatment is the heating of metals to or above certain critical temperatures followed by controlled cooling at critical speeds in order to modify the properties of the metal or alloy. Fast cooling is referred to as quenching. The temperatures and rates of cooling vary from metal to metal and from alloy to alloy. Medium- and high-carbon steels can be 'quench hardened' by heating to a specified temperature and cooling them quickly by plunging the heated workpiece into a tank of cold water. These metals can be softened ('annealed') by heating them to the same temperature but cooling them very slowly.

Most copper alloys can only be work hardened by cold-working processes such as rolling which distorts their grain structure. They can be softened (annealed) by heating. In this case the rate of cooling is not important.

Some aluminium alloys can also be heat treated. They can be 'precipitation hardened' and they can be softened by 'solution treatment'.

6.2.5 *Hot and cold working*

The terms hot working and cold working are used from time to time in this chapter and elsewhere in this book and in other books in this series. We need to know about hot and cold working because these processes change the properties of metals and alloys.

Metals may be shaped by making them molten and casting them to shape in moulds. Metals may also be shaped by cutting them with hand and machine tools. However, there is the alternative of *working* them to shape. This is how a blacksmith shapes metal by hammering it to shape on the anvil. Metal that has been shaped by working (also called 'flow forming') is said to be *wrought*.

Metals may be worked hot or cold. In both cases the grain of the metal is distorted by the process and the metal becomes harder, stronger and less ductile.

- In *cold-worked* metals the grain becomes distorted due to the processing and remains distorted when the processing has finished. This leaves the metal harder, stronger and less ductile. Unfortunately, further cold working could cause the metal to crack.
- In *hot-worked* metals the grain also becomes distorted but the metal is sufficiently hot for the grains to reform as fast as the distortion occurs, leaving the metal soft and ductile. Some grain refinement will occur so that the metal should be stronger than at the start of the process. The metal is easier to work at high temperatures and less force is required to form it. This is why the blacksmith makes his metal red hot before forging it to shape with a hammer.
- *Recrystallisation* is the term used when the distorted grains reform when they are heated. The temperature at which this happens depends upon the type of metal or alloy and how severely it has been processed by cold working.
- *Cold working* is the flow forming of metal *below* the temperature of recrystallisation. For example, cold heading rivets.
- *Hot working* is the flow forming of metals *above* the temperature of recrystallisation. For example, rolling red-hot ingots into girders in a steel mill.

Figure 6.5 shows examples of hot and cold working, and Tables 6.1 and 6.2 summarise the advantages and limitations of hot- and cold-working processes.

Fig. 6.5 *Examples of (a) hot-working and (b) cold-working processes*

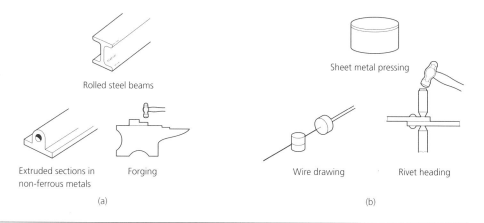

Rolled steel beams

Extruded sections in non-ferrous metals

Forging

(a)

Sheet metal pressing

Wire drawing

Rivet heading

(b)

Table 6.1 *Hot-working processes*

Advantages	Limitations
1. Low cost	1. Poor surface finish – rough and scaly
2. Grain refinement from cast structure	2. Due to shrinkage on cooling the dimensional accuracy of hot-worked components is of a low order
3. Materials are left in the fully annealed condition and are suitable for cold working (heading, bending, etc.)	3. Due to distortion on cooling and to the processes involved, hot working generally leads to geometrical inaccuracy
4. Scale gives some protection against corrosion during storage	4. Fully annealed condition of the material coupled with a relatively coarse grain leads to a poor finish when machined
5. Availability as sections (girders) and forgings as well as the more usual bars, rods, sheets, strip and butt-welded tube	5. Low strength and rigidity for metal considered
	6. Damage to tooling from abrasive scale on metal surface

Table 6.2 *Cold-working processes*

Advantages	Limitations
1. Good surface finish	1. Higher cost than for hot-worked materials. It is only a finishing process for material previously hot worked. Therefore, the processing cost is added to the hot-worked cost
2. Relatively high dimensional accuracy	2. Materials lack ductility due to work hardening and are less suitable for bending, etc.
3. Relatively high geometrical accuracy	3. Clean surface is easily corroded
4. Work hardening caused during the cold-working processes: (a) increases strength and rigidity (b) improves the machining characteristics of the metal so that a good finish is more easily achieved	4. Availability limited to rods and bars also sheets and strip, solid drawn tubes

6.3 Classification of materials

Almost every known substance has found its way into the engineering workshop at some time. Neither this chapter nor, in fact, this book could hold all the facts about all the materials used by engineers. Therefore, to keep things simple, we will first group these materials into similar types and then consider the properties and uses of some examples from each group. These main groups are shown in Fig. 6.6.

Fig. 6.6 *Classification of engineering materials*

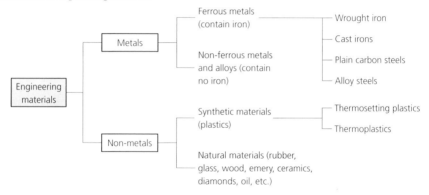

6.3.1 *Metals*

Of the 93 known elements, some 70 are classified as metals. It is difficult to define metals without resorting to some advanced scientific theory. However, at this level, we can consider metals as substances that have a lustrous sheen when cut, are good conductors of heat and are good conductors of electricity. Some examples are aluminium, copper and iron. Sometimes metals are mixed with non-metals – for example, cast irons and plain carbon steels are mixtures of iron and carbon with traces of other elements. Sometimes metals are mixed with other metals to form *alloys* – for example, brass is an alloy of copper and zinc.

6.3.2 *Non-metals*

These are all the substances that are not metals. They can be elements, compounds of elements and mixtures of compounds. They include wood, rubber, plastics, ceramics and glass. Some materials are compounds of metals and non-metals. For example, naturally occurring abrasive grits, such as emery and corundum contain between 70 and 90 per cent of aluminium oxide (a compound of aluminium and oxygen). Aluminium oxide (also known as alumina) is used in firebricks for furnace linings.

Organic compounds are based on the element *carbon* chemically combined with other substances. Some examples of organic materials are natural materials such as wood and some rubbers, and synthetic materials such as plastics.

6.4 Ferrous Metals (wrought iron and plain carbon steels)

Ferrous metals and alloys are based on the metal iron which is their main constituent. They are called ferrous metals because the Latin name for iron is *Ferrum*. Iron is a soft grey metal and is rarely found in the pure state outside a laboratory. For engineering purposes the metal iron is usually associated with the non-metal carbon.

6.4.1 *Wrought iron*

Wrought iron is seldom used nowadays. It is a ferrous metal with such a low carbon content that the iron–carbon compounds essential to a steel cannot be formed. Also a wrought iron bar contains fibres of slag running through it. These fibres of slag are left over from the process by which wrought iron is made; they improve its corrosion resistance and prevent sudden fractures occurring. The metal tends to tear rather than snap suddenly. Figure 6.7 outlines some properties and typical applications for wrought iron.

Fig. 6.7 *Wrought iron: properties and uses*

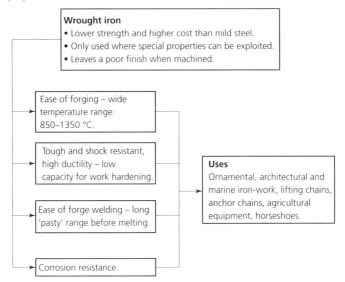

6.4.2 *Plain carbon steels*

Plain carbon steels consist, as their name implies, mainly of iron with small quantities of carbon. There will also be traces of impurities left over from when the metallic iron was extracted for its mineral ore. A small amount of the metal manganese is added to counteract the effects of the impurities. However, the amount of manganese present is insufficient to change the properties of the steel and it is, therefore, not considered to be an alloying element. Plain carbon steels may contain:

0.1–1.4 per cent carbon
up to 1.0 per cent manganese (not to be confused with magnesium)
up to 0.3 per cent silicon
up to 0.05 per cent sulphur
up to 0.05 per cent phosphorus

Figure 6.8 shows how the carbon content of a plain carbon steel affects the properties of the steel. For convenience we can group plain carbon steels into three categories:

- low-carbon steels (below 0.3 per cent carbon)
- medium-carbon steels (0.3–0.8 per cent carbon)
- high-carbon steels (between 0.8 and 1.4 per cent carbon)

Fig. 6.8 *The effect of carbon content on the properties of plain carbon steels (annealed)*

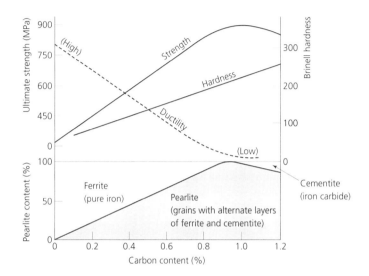

Low-carbon steels

These are also referred to as mild steels. If the carbon content is kept between 0.1 and 0.15 per cent the steel is often referred to as 'dead mild' steel. This steel is very ductile and very soft and can be pressed into complicated shapes for car body panels at room temperature without cracking. It is not used for machined components since its softness would cause it to tear and leave a poor surface finish. It is slightly weaker than the next group of low-carbon steels to be considered.

If the carbon content is between 0.15 and 0.3 per cent the steel is stronger, but slightly less soft and ductile. It is often referred to as 'mild' steel. It can be forged, rolled and drawn both in the hot and in the cold condition. It is easily machined with high-speed steel-cutting tools. Because of its ease of manufacture and the very large quantities produced, mild steels are the cheapest and most plentiful of the steel products. Figure 6.9 outlines some properties and uses of low-carbon steels.

Some low-carbon steel contain additives such as lead or sulphur to make them 'free-cutting'. The greatly improved machining properties is only achieved at the expense of the strength and toughness of the steels.

Fig. 6.9 *Low-carbon steel: properties and uses*

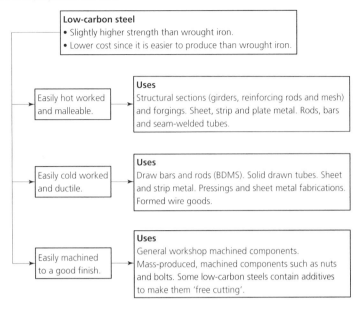

Low-carbon steel
- Slightly higher strength than wrought iron.
- Lower cost since it is easier to produce than wrought iron.

| Easily hot worked and malleable. | **Uses** Structural sections (girders, reinforcing rods and mesh) and forgings. Sheet, strip and plate metal. Rods, bars and seam-welded tubes. |

| Easily cold worked and ductile. | **Uses** Draw bars and rods (BDMS). Solid drawn tubes. Sheet and strip metal. Pressings and sheet metal fabrications. Formed wire goods. |

| Easily machined to a good finish. | **Uses** General workshop machined components. Mass-produced, machined components such as nuts and bolts. Some low-carbon steels contain additives to make them 'free cutting'. |

Medium-carbon steels

There are two groups of medium-carbon steels:

- 0.3–0.5 per cent carbon. These can be heat treated to make them tough and strong.
- 0.5–0.8 per cent carbon. These can be heat treated to make them fairly hard yet remain impact resistant.

Medium-carbon steels are harder, stronger and tougher than low-carbon steels. They are also more expensive. They cannot be bent or formed in the cold condition to the same extent as low-carbon steels without cracking. However, medium-carbon steels hot forge well, but close temperature control is required to prevent:

- 'Burning' at high temperatures over 1150 °C, as this leads to embrittlement. The metal cannot be reclaimed and the forging has to be scrapped.
- Cracking when forging is continued below 700 °C. This is due to work hardening as the steel is in the 'cold' condition from a forging point of view.

Figure 6.10 outlines some properties and uses for medium-carbon steels. For many of these applications the steel has to be heat treated to enhance its properties. The heat treatment of medium-carbon steels is considered in Chapter 7.

High-carbon steels

These are harder, less ductile and more expensive than both mild- and medium-carbon steels. They are also less tough and are mostly used for springs, cutting tools and forming dies. High-carbon steels work harden readily and, for this reason, they are not recommended for cold working. However, they forge well providing the temperature is controlled at between 700 and 900 °C. There are three groups of high-carbon steels:

Fig. 6.10 *Medium-carbon steel: properties and uses*

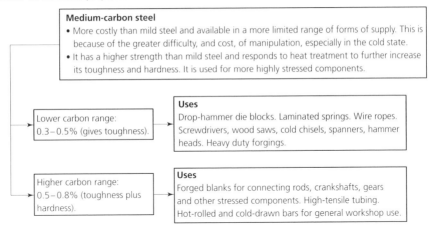

Medium-carbon steel
- More costly than mild steel and available in a more limited range of forms of supply. This is because of the greater difficulty, and cost, of manipulation, especially in the cold state.
- It has a higher strength than mild steel and responds to heat treatment to further increase its toughness and hardness. It is used for more highly stressed components.

Lower carbon range: 0.3–0.5% (gives toughness).

Uses
Drop-hammer die blocks. Laminated springs. Wire ropes. Screwdrivers, wood saws, cold chisels, spanners, hammer heads. Heavy duty forgings.

Higher carbon range: 0.5–0.8% (toughness plus hardness).

Uses
Forged blanks for connecting rods, crankshafts, gears and other stressed components. High-tensile tubing. Hot-rolled and cold-drawn bars for general workshop use.

- 0.8–1.0 per cent carbon where both toughness and hardness are required. For example, cold chisels, hammers heads and axes heads.
- 1.0–1.2 per cent carbon for sufficient hardness for most metal cutting tools.
- 1.2–1.4 per cent where extreme hardness is required for wood-working tools and knives that need a very keen cutting edge.

Figure 6.11 outlines some properties and uses for high-carbon steels. For all of these applications the steel has to be heat treated to enhance its properties. The heat treatment of high-carbon steels is considered in Chapter 7.

Fig. 6.11 *High-carbon steel: properties and uses*

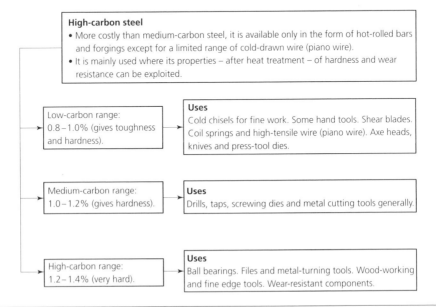

High-carbon steel
- More costly than medium-carbon steel, it is available only in the form of hot-rolled bars and forgings except for a limited range of cold-drawn wire (piano wire).
- It is mainly used where its properties – after heat treatment – of hardness and wear resistance can be exploited.

Low-carbon range: 0.8–1.0% (gives toughness and hardness).

Uses
Cold chisels for fine work. Some hand tools. Shear blades. Coil springs and high-tensile wire (piano wire). Axe heads, knives and press-tool dies.

Medium-carbon range: 1.0–1.2% (gives hardness).

Uses
Drills, taps, screwing dies and metal cutting tools generally.

High-carbon range: 1.2–1.4% (very hard).

Uses
Ball bearings. Files and metal-turning tools. Wood-working and fine edge tools. Wear-resistant components.

6.5 Ferrous metals (alloy steels)

These are essentially plain carbon steels to which other metals (*alloying elements*) have been added in sufficient quantities to materially alter the properties of the steel. The most common alloying elements are:

- *Nickel*, which refines the grain and strengthens the steel.
- *Chromium*, which improves the response of the steel to heat treatment; it also improves the corrosion resistance of the steel.
- *Molybdenum*, which reduces temper brittleness during heat treatment, welding, and operation at sustained high temperatures.
- *Manganese*, which improves the strength and wear resistance of steels. Steels containing a high percentage of manganese (14 per cent) are highly wear resistant and these steels are used for such applications as bulldozer blades and plough blades.
- *Tungsten* and *cobalt*, which improve the ability of a steel to remain hard at high temperatures and are used extensively in cutting tool materials.

Alloy steels are used where it is necessary to have great strength; where corrosion resistance is required, and where the ability to remain hard at high temperatures is required. In this book we only need to consider high-speed steels for cutting tools, and stainless steels where corrosion resistance is required.

6.5.1 *High-speed steels*

High-speed steels are alloy steels. Figure 6.12 shows some typical high-speed steel-cutting tools suitable for cutting metals. These cutting tools are for use with machine tools where the heat generated by the cutting process would soon soften high-carbon steel tools. High-speed steels can operate continuously at 700 °C, whereas high-carbon steel starts to soften at 220 °C.

Figure 6.13 outlines the availability and properties of typical high-speed steels, and Table 6.3 lists the composition and uses of these steels.

6.5.2 *Stainless steels*

These are also alloy steels. They contain a high proportion of chromium to provide corrosion resistance. Various grades of stainless steel and their applications are listed in Table 6.4. Because it is ductile and so easily formed, 18/8 stainless steel (BS 302S25) is the most widely used alloy. Your kitchen sink is most likely made of this alloy. Unfortunately stainless steel fabrications tend to suffer from *weld decay*. If it is known that stainless steel components are to be assembled by welding, then the BS 321S20 alloy should be used. This has been *proofed* by the inclusion of titanium and does not suffer from weld decay.

Fig. 6.12 *Some typical high-speed steel cutting tools: (a) milling cutters; (b) twist drills; (c) butt-welded turning tools; (d) hacksaw blades*

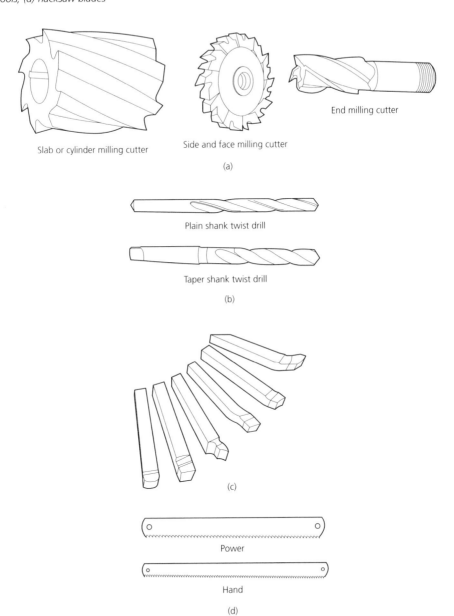

End milling cutter

Slab or cylinder milling cutter

Side and face milling cutter

(a)

Plain shank twist drill

Taper shank twist drill

(b)

(c)

Power

Hand

(d)

Fig. 6.13 *High-speed steel: properties and uses*

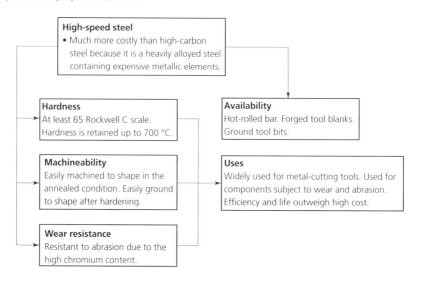

High-speed steel
• Much more costly than high-carbon steel because it is a heavily alloyed steel containing expensive metallic elements.

Hardness
At least 65 Rockwell C scale. Hardness is retained up to 700 °C.

Availability
Hot-rolled bar. Forged tool blanks. Ground tool bits.

Machineability
Easily machined to shape in the annealed condition. Easily ground to shape after hardening.

Uses
Widely used for metal-cutting tools. Used for components subject to wear and abrasion. Efficiency and life outweigh high cost.

Wear resistance
Resistant to abrasion due to the high chromium content.

Table 6.3 *Typical high-speed steels*

Type of steel	Composition* (%)						Hardness (VNP)	Uses
	C	Cr	W	V	Mo	Co		
18% tungsten	0.68	4.0	19.0	1.5	—	—	800–850	Low-quality alloy, not much used
30% tungsten	0.75	4.7	22.0	1.4	—	—	850–950	General-purpose cutting tools for jobbing work shops
6% cobalt	0.8	5.0	19.0	1.5	0.5	6.0	800–900	Heavy-duty cutting tools
Super HSS 12% cobalt	0.8	5.0	21.0	1.5	0.5	11.5	850–950	Heavy-duty cutting tools for machining high-tensile materials

*C = Carbon, Cr = chromium, W = tungsten, V = vanadium, Mo = molybdenum, Co = cobalt.

Table 6.4 Typical stainless steels

Type of steel	Composition					Mechanical properties				Heat treatment	Applications
	C	Mn	Cr	Ni	Si	R_m	R_e	A	H_B		
403S17 Ferritic	0.04	0.45	14.0	0.50	0.80	510	340	31	—	Condition soft. Cannot be hardened except by cold work	Soft and ductile; can be used for fabrications, pressings, drawn components, spun components. Domestic utensils
420S45 Martensitic	0.30	1.0	13.0	1.0	0.80	1470 1670	— —	— —	450 534	Quench from 950–1000 °C. Temper 400–450 °C. Temper 150–180 °C	Corrosion-resistant springs for food processing and chemical plant. Corrosion-resistant cutlery and edge tools
302S25 Austenitic	0.1	1.0	18.0	8.50	0.80	61 896	278 803	50 30	170 —	Condition soft solution treatment from 1050 °C	18/8 stainless steel widely used for fabrications and domestic and decorative purposes

6.6 Ferrous metal (cast irons)

These are also ferrous metals containing iron and carbon. They do not require the expensive refinement processes of steel making and provide a relatively low-cost engineering material that can easily be cast into complex shapes at much lower temperatures that those associated with cast steel. There are four main groups of cast iron:

- Grey cast iron.
- Spheroidal graphite cast iron.
- Malleable cast iron.
- Alloy cast ion.

In this book we are only concerned with *grey cast iron*. Unlike steels, where the carbon content is deliberately restricted so that it can all combine with the iron, in cast irons there is so much carbon present (about 3 per cent) that not all of it can combine with the iron and some of the carbon is left over. This surplus carbon appears as flakes of graphite between the crystal of metal, as shown Fig. 6.14(a).

Graphite is the type of carbon that is used to make pencil leads and it is these flakes of graphite that give cast irons their characteristic grey colour when fractured, its 'dirtiness' when machined or filed, and its weakness when subjected to a tensile load. The graphite also promotes good machining characteristics by acting as an internal lubricant and causes the chips (swarf) to break up into granules that can easily be disposed of. The cavities containing the flakes of graphite have a damping effect upon vibrations – cast iron is *anti-resonant* – and this property makes it particularly suitable for machine tool frames and beds. The graphite also tends to make the slideways self-lubricating. Figure 6.14(b) shows a typical cast-iron machine tool component. The properties and uses of grey cast iron are outlined in Fig. 6.15.

Fig. 6.14 *Structure and an application of cast iron: (a) grey cast iron; (b) typical casting*

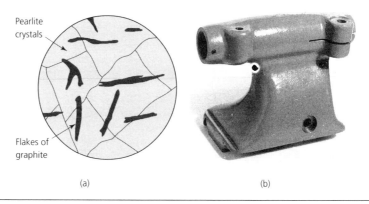

Pearlite
crystals

Flakes of
graphite

(a)   (b)

Fig. 6.15 *Grey cast iron: properties and uses*

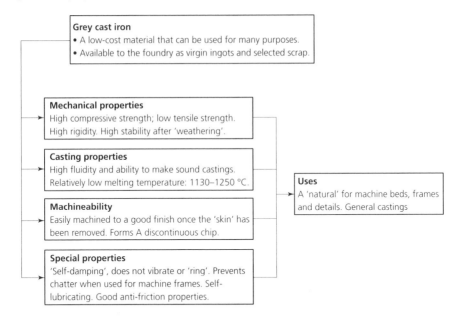

Grey cast iron
- A low-cost material that can be used for many purposes.
- Available to the foundry as virgin ingots and selected scrap.

Mechanical properties
High compressive strength; low tensile strength.
High rigidity. High stability after 'weathering'.

Casting properties
High fluidity and ability to make sound castings.
Relatively low melting temperature: 1130–1250 °C.

Machineability
Easily machined to a good finish once the 'skin' has
been removed. Forms A discontinuous chip.

Special properties
'Self-damping', does not vibrate or 'ring'. Prevents
chatter when used for machine frames. Self-
lubricating. Good anti-friction properties.

Uses
A 'natural' for machine beds, frames
and details. General castings

6.7 Non-ferrous metals and alloys

Non-ferrous metals and alloys refer to all the multitude of metals and alloys that do not
contain iron or, if any iron is present, it is only a minute trace. The most widely used non-
ferrous metals and alloys are:

- Aluminium and its alloys.
- Copper and its alloys.
- Zinc-based die-casting alloys.
- Titanium and its alloys used in aerospace engineering, including airframe and engine
  components.

In this book we are only interested in the first two groups.

6.7.1 *Aluminium and its alloys*

This is the lightest of the commonly used metals. Its electrical and thermal conductivity
properties are very good, being second only to copper. It also has good corrosion
resistance and is cheaper than copper. Unfortunately, it is relatively mechanically weak
in the pure state and is difficult to solder and weld. Special techniques and materials are

required for these processes. The properties and uses of pure aluminium are outlined in Fig. 6.16

Fig. 6.16 *Aluminium: properties, uses and forms of supply*

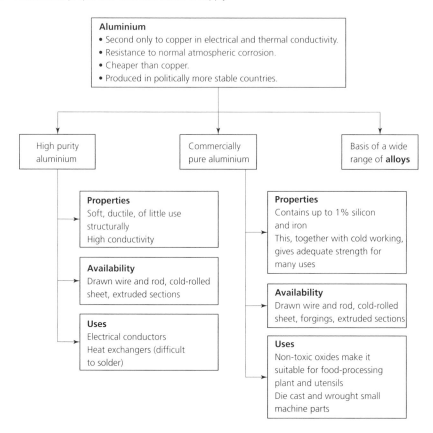

Because of its high electrical conductivity, and light weight, aluminium conductors are used for the overhead grid system. The cable is made up of aluminium conductors surrounding a core made up from high-tensile steel wire. The steel supports the mechanical loads and the aluminium conducts the electricity and protects the core from atmospheric corrosion.

Aluminium is also the basis of a wide range of alloys. These can be classified as:

- Wrought alloys (not heat treatable).
- Wrought alloys (heat treatable).
- Casting alloys (not heat treatable).
- Casting alloys (heat treatable).

The composition and uses of some typical examples of each group of aluminium alloys are listed in Table 6.5.

6.7.2 *Copper and its alloys*

Copper has already been introduced as a corrosion-resistant metal with excellent electrical and thermal conductivity properties. It is also relatively strong compared with aluminium and very easy to join by soldering or brazing. It is very much heavier than aluminium and also more costly. There are various grades of copper and the use and properties of these are outlined in Fig. 6.17.

Fig. 6.17 *Copper: properties, uses and forms of supply*

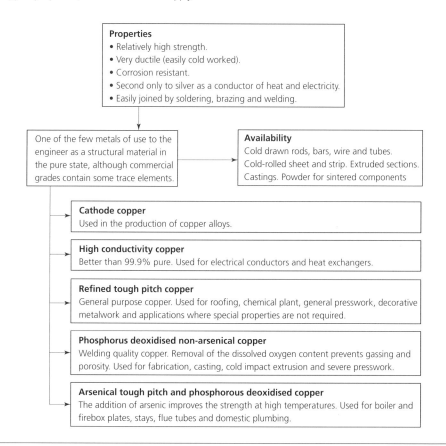

Table 6.5 Some typical aluminium alloys

Composition (%)

Copper	Silicon	Iron	Manganese	Magnesium	Other elements	Category	Applications
0.1 max.	0.5 max.	0.7 max.	0.1 max.	—	—	Wrought Not heat-treatable	Fabricated assemblies. Electrical conductors. Food and brewing processing plant. Architectural decoration
0.15 max.	0.6 max.	0.75 max.	1.0 max.	4.5–5.5	0.5 chromium	Wrought Not heat-treatable	High-strength shipbuilding and engineering products. Good corrosion resistance
1.6	10.0	—	—	—	—	Cast Not heat-treatable	General purpose alloy for moderately stressed pressure die-castings
—	10.0–13.0	—	—	—	—	Cast Not heat-treatable	One of the most widely used alloys. Suitable for sand, gravity and pressure die-castings. Excellent foundry characteristics for large marine, automative and general engineering castings

4.2	0.7	0.7	0.7	0.7	0.3 titanium (optional)	Wrought Heat-treatable	Traditional 'Duralumin' general machining alloy. Widely used for stressed components in aircraft and elsewhere
—	0.5	—	—	0.6	—	Wrought Heat-treatable	Corrosion-resistant alloy for lightly stressed components such as glazing bars, window sections and automotive body components
1.8	2.5	1.0	—	0.2	0.15 titanium 1.2 nickel	Cast Heat-treatable	Suitable for sand and gravity die-casting. High rigidity with moderate strength and shock resistance. A general purpose alloy
—	—	—	—	10.5	0.2 titanium	Cast Heat-treatable	A strong, ductile and highly corrosion-resistant alloy used for aircraft and marine castings both large and small

The main groups of copper-based alloys are:

- The brass alloys (copper and zinc).
- The tin–bronze alloys (copper and tin).
- The aluminium–bronze alloys (copper and aluminium).
- The cupro-nickel alloys (copper and nickel).

In this book we are only interested in the first two groups, namely the brass alloys and the tin–bronze alloys.

Brass alloys

Brass alloys of copper and zinc tend to give rather weak and porous castings. The brasses depend largely upon hot and/or cold working to consolidate the metal and improve its mechanical properties. The brass alloys can only be hardened by cold working (work hardening). They can be softened by heat treatment (the annealing process). The composition and uses of the more commonly available brass alloys are given in Table 6.6.

Table 6.6 *Typical brass alloys*

Name	Composition (%) Copper	Zinc	Other elements	Applications
Cartridge brass	70	30	—	Most ductile of the copper–zinc alloys. Widely used in sheet metal pressing for severe deep drawing operations. Originally developed for making cartridge cases, hence its name
Standard brass	65	35	—	Cheaper than cartridge brass and rather less ductile. Suitable for most engineering processes
Basis brass	63	37	—	The cheapest of the cold-working brasses. It lacks ductility and is only capable of withstanding simple forming operations
Muntz metal	60	40	—	Not suitable for cold working, but hot works well. Relatively cheap due to its high zinc content, it is widely used for extrusion and hot-stamping processes
Free-cutting brass	58	39	3% lead	Not suitable for cold working, but excellent for hot working and high-speed machining of low-strength components
Admiralty brass	70	29	1% tin	This is virtually cartridge brass plus a little tin to prevent corrosion in the presence of salt water
Naval brass	62	36	1% tin	This is virtually Muntz metal plus a little tin to prevent corrosion in the presence of salt water

Tin–bronze alloys

These are alloys of copper and tin together with a *deoxidiser*. The deoxidiser is essential to prevent the tin content from oxidising at high temperatures during casting and hot working. Oxidation is the chemical combination of the tin content with the oxygen in the atmosphere, and results in the bronze being weakened and becoming hard, brittle and 'scratchy'. Two deoxidisers are commonly used:

- A small amount of phosphorus in the *phosphor bronze* alloys.
- A small amount of zinc in the *gun metal* alloys.

The composition and uses of some typical tin–bronzes are listed in Table 6.7. Unlike the brasses, which are largely used in the wrought condition (rod, sheet, etc.), only low tin-content bronzes can be worked and most bronze components are in the form of castings. Tin–bronzes are more expensive than brasses, but are stronger, and give sound, pressure-tight castings that are widely used for steam and hydraulic valve bodies and mechanisms. They are highly resistant to corrosion.

6.8 Workshop tests for the identification of metals

Materials represent a substantial investment in any manufacturing company. It is essential that all materials are carefully stored so that they are not damaged or allowed to deteriorate before use. Ferrous metals must be stored in a warm, dry environment so that rusting cannot occur. This is particularly the case when storing bright drawn sections and centreless ground rod. Rusting would destroy the finish and cause such materials to be unfit for use. Materials must also be carefully labelled or coded so that they can be quickly and accurately identified. Mistakes, resulting in the use of the incorrect material, can be very costly through waste, lost man-hours and machine time.

The similarity in appearance between many metals of different physical properties makes it essential that some form of permanent identification should be marked on them, e.g. colour coding. However, mix-ups do occur from time to time and, also, bar 'ends' are often used up for 'one-off' jobs. Table 6.8 gives some simple workshop tests of identification.

6.9 Non-metals (natural)

Non-metals are widely used in engineering today. Some of the materials occur naturally. For example:

- *Rubber* is used for anti-vibration mountings, coolant and compressed air hoses, transmission belts, truck wheel tyres.
- *Glass* is used for spirit level vials (the tube that contains the bubble), lenses for optical instruments.
- *Emery and corundum* (aluminium oxides) are used abrasive wheels belts and sheets, and as grinding pastes. Nowadays they are usually produced artificially to control the quality.
- *Wood* is used for making casting patterns.
- *Ceramics* are used for cutting tool tips and electrical insulators.

Table 6.7 Typical tin–bronze alloys

Name	Composition (%)					Application
	Copper	Zinc	Phosphorus	Tin	Lead	
Low-tin bronze	96	—	0.1–0.25	3.9–3.75	—	This alloy can be severely cold worked to harden it so that it can be used for springs where good elastic properties must be combined with corrosion resistance, fatigues resistance and electrical conductivity, e.g. contact blades
Drawn phosphor-bronze	94	—	0.1–0.5	5.9–5.5	—	This alloy is used in the work-hardened condition for turned components requiring strength and corrosion resistance, such as valve spindles
Cast phosphor-bronze	rem.	—	0.03–0.25	10	—	Usually cast into rods and tubes for making bearing bushes and worm wheels. It has excellent anti-friction properties
Admiralty gunmetal	88	2	—	10	—	This alloy is suitable for sand casting where fine-grained, pressure-tight components such as pump and valve bodies are required
Leaded-gunmetal (free-cutting)	85	5	—	5	5	Also known as 'red brass', this alloy is used for the same purposes as standard, Admiralty gunmetal. It is rather less strong but has improved pressure tightness and machine properties
Leaded (plastic) bronze	74	—	—	2	24	This alloy is used for lightly loaded bearings where alignment is difficult. Due to its softness, bearings made from this alloy 'bed in' easily

Table 6.8 *Workshop identification tests*

These are not foolproof and require some experience

Metal	Appearance	Hammer cold	Type of chip	'Spark test' on grinding wheel
Mild steel ('black')	Smooth scale with blue/black sheen	Flattens easily	Smooth, curly ribbon-like	Stream of yellow white sparks, varying in length: slightly 'fiery'
Mild steel ('bright')	Smooth, scale-free, silver grey surface	Flattens easily	Smooth, curly ribbon-like	Stream of yellow white sparks, varying in length: slightly 'fiery'
Medium-carbon steel	Smooth scale, black sheen	Fairly difficult to flatten	Chip curls more tightly and discolours light brown	Yellow sparks, shorter than m/s, and finer and more feathery
High-carbon steel	Rougher scale, black	Difficult to flatten	Chip curls even more tightly and discolours dark blue	Sparks less bright, starting near grinding wheel, and more feathery with secondary branching (distinctive acrid smell)
High-speed steel	Rougher scale, black with reddish tint	Very difficult to flatten. Tends to crack easily	Long ribbon-like chip. Distinctive smell. Over-heats tool easily	Faint red streak ending in fork (distinctive acrid smell)
Cast iron	Grey and sandy	Crumbles	Granular, grey in colour	Faint red spark, ending in bushy yellow sparks (distinctive acrid smell)
Copper	Distinctive 'red' colour	Flattens very easily	Ribbon-like, with razor edge	Should not be ground – no sparking
Aluminium	Silvery when polished. Pale grey when oxidised. Very light in weight compared with other metals	Flattens very easily	Ribbon-like chip	Should not be ground – no sparking

6.10 Non-metals (synthetic)

These are popularly known as *plastics*. When we were considering the properties of materials, a plastic material is said to be one which deforms to a new shape under an applied load and retains its new shape when the load is removed. Yet, the range of synthetic materials we call *plastics* are often tough and leathery, or hard and brittle, or elastic. They are called *plastics* because, during the moulding operation by which they are formed, they are reduced to a plastic condition by heating them to about twice the temperature of boiling water.

There are many families of 'plastic' materials with widely differing properties. However, they all have certain properties in common.

- *Electrical insulation* All plastic materials are, to a greater or lesser extent, good electrical insulators (they are also good heat insulators). However, their usefulness as insulators is limited by their inability to withstand high temperatures and their relative softness compared with ceramics. They are mainly used for insulating wires and cables and for moulded switch gear and instrument components and cases.
- *Strength/weight ratio* Plastic materials vary considerably in strength. All plastics are much less dense than metals and this results in a favourable strength/weight ratio. The high-strength plastics and reinforced plastics compare favourably with the aluminium alloys and are often used for stressed components in aircraft construction.
- *Degradation* Plastics do not corrode like metals. They are all inert to most inorganic chemicals. They can be used in environments that are chemically hostile to even the most corrosion-resistant metals. They are superior to natural rubber in their resistance to attack by oils and greases. However, all plastics degrade at high temperatures and many are degraded by the ultraviolet content of sunlight. Plastics that have to be exposed to sunlight (window frames and roof guttering) usually contain a pigment that filters out the ultraviolet rays. Some thermoplastics can be dissolved by suitable solvents.
- *Safety* Plastics can give off very dangerous toxic fumes when heated. Note the number of people who have died from inhaling the smoke from plastic furniture padding in house fires! Solvents used in the processing of plastics are often highly toxic and should not be inhaled but used in well-ventilated surroundings. Make sure you know the likely dangers before starting work on plastic materials and always follow the safe working practices laid down by the safety management.

Plastic materials can be grouped into two distinct families. These are the *thermosetting plastics* and the *thermoplastics*. Typical examples of each of these families will now be considered. Note that thermosetting plastics are often referred to simply as 'thermosets'.

6.10.1 *Thermosetting plastics*

These undergo a chemical change during moulding. This is called 'curing'. It is caused by the heating of the plastic material (powder or granules) during moulding. Once this chemical change has taken place, and the moulding is said to be 'cured', the plastic material from which the moulding is made can never again be softened and reduced to a plastic condition by reheating.

Thermosetting resins are unsuitable for use by themselves and they are usually mixed with other substances (additives) to improve their mechanical properties, improve their moulding properties, make them more economical to use, and provide the required colour for the finished product. A typical moulding material could consist of:

Resin	38 per cent by weight
Filler	58 per cent by weight
Pigment	3 per cent by weight
Mould release agent	0.5 per cent by weight
Catalyst	0.3 per cent by weight
Accelerator	0.2 per cent by weight

The pigment gives colour to the finished product. The mould release agent stops the moulding sticking to the mould. It also acts as an internal lubricant and helps the plasticised material to flow to the shape of the mould. The catalyst promotes the curing process and the accelerator speeds up the curing process and reduces the time the moulds have to be kept closed, thus improving productivity.

Fillers are much cheaper than the resin itself and this is important in keeping down the cost of the moulding. Fillers also have a considerable influence on the properties of the mouldings produced from a given thermosetting resin. They improve the impact strength (toughness) and reduce shrinkage during moulding. The following are some typical fillers:

- *Shredded paper* and *shredded cloth* give good strength and reasonable electrical insulation properties at a low cost.
- *Mica granules* give good strength and heat resistance (asbestos is no longer used).
- *Aluminium powder* gives good mechanical strength and wear resistance.
- *Wood flour (fine sawdust)* and *calcium carbonate (ground limestone)* provide high bulk at a very low cost but with relatively low strength.
- *Glass fibre (chopped)* gives good strength and excellent electrical insulation properties.

Some typical examples of thermosetting plastics and their uses are given in Table 6.9.

Table 6.9 *Some typical thermosetting plastic materials*

Material	Characteristics
Phenolic resins and powders	These are used for dark-coloured parts because the basic resin tends to become discoloured. These are heat-curing materials
Amino (containing nitrogen) resins and powders	These are colourless and can be coloured if required; they can be strengthened by using paper-pulp fillers, and used in thin sections
Polyester resins	Polyester chains can be cross-linked by using a monomer such as styrene; these resins are used in the production of glass-fibre laminates
Epoxy resins	These are also used in the production of glass-fibre laminates

6.10.2 *Thermoplastics*

These can be softened as often as they are reheated. They are not as rigid as the thermosetting plastics but they tend to be tougher. Additives (other than a colourant and an internal lubricant) are not normally used with thermoplastics. Some typical examples of thermoplastics and their uses are given in Table 6.10.

Table 6.10 *Some typical thermoplastic materials*

Type	Material	Characteristics
Cellulose plastics	Nitrocellulose	Materials of the 'celluloid' type are tough and water resistant. They are available in all forms except moulding powders. They cannot be moulded because of their flammability
	Cellulose acetate	This is much less flammable than the above. It is used for tool handles and electrical goods
Vinyl plastics	Polythene	This is a simple material that is weak, easy to mould, and has good electrical properties. It is used for insulation and for packaging
	Polypropylene	This is rather more complicated than polythene and has better strength
	Polystyrene	Polystyrene is cheap, and can be easily moulded. It has a good strength but it is rigid and brittle and crazes and yellows with age
	Polyvinyl chloride (PVC)	This is tough, rubbery, and almost non-inflammable. It is cheap and can be easily manipulated: it has good electrical properties
Acrylics (made from an acrylic acid)	Polymethyl methacrylate	Materials of the 'Perspex' type have excellent light transmission, are tough and non-splintering, and can be easily bent and shaped
Polyamides (short carbon chains that are connected by amide groups – NHCO)	Nylon	This is used as a fibre or as a wax-like moulding material. It is fluid at moulding temperature, tough, and has a low coefficient of friction
Fluorine plastics	Polytetrafluoroethylene (PTFE)	Is a wax-like moulding material; it has an extremely low coefficient of friction. It is very expensive
Polyesters (when an alcohol combines with an acid, an 'ester' is produced)	Polyethylene terephthalate	This is available as a film or as 'Terylene'. The film is an excellent electrical insulator

6.11 Forms of supply

There is an almost unlimited range to the forms of supply in which engineering materials can be supplied to a manufacturer or to a fabricator. Figure 6.18 shows some of these forms of supply.

Fig. 6.18 *Forms of supply*

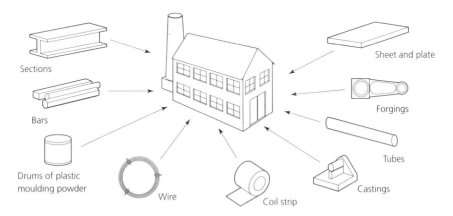

- *Sections* For steel these can be steel angles, channel sections, H-section beams and joists and T-sections in a wide range of sizes and lengths. They are usually hot rolled with a heavily scaled finish. Such sections are mostly used in the steel fabrication and civil engineering and construction industries. Bright-drawn steel angle sections are available in the smallest sizes. Non-ferrous metal sections are normally extruded to close tolerances and have a bright finish. Both standard sections and sections to customers' own requirements are made this way. The sizes available are very much smaller than for steel sections.
- *Bars* These may be 'flats' which are available in rectangular sections, they may be 'squares', or they may be 'rounds' which have a cylindrical section, and this term applies to the larger sizes. In the smaller size they are usually referred to as 'rods'. They are available in hot-rolled (black) or cold-drawn (bright) finishes. Hexagon section bars for making nuts and bolts are always cold drawn.
- *Plastics* These may be supplied in powder or granular form for moulding into various shapes. They may also be supplied semi-finished in rounds, flats, squares, tubes and sheet. They may also come in sections to manufacturers' own requirements such as curtain rails and insulation blocks ready for cutting off to length. Reinforced plastics, such as 'Tufnol', are also available in standard sections and mouldings.

- *Wire*   This is available bright drawn or hot rolled, depending upon its size and the use to which it is going to be put. Bright-drawn, high-carbon steel wire (piano wire) is used for making springs. Copper wire is used for electrical conductors. The smaller the diameter of any wire, the longer the length that can be made.
- *Coil strip*   This is used for cold stamping and pressing where continuous automatic feed to the presses is required. It is available in a range of thickness and finishes. Steel strip is available in such finishes as bright rolled (BR), hot rolled, pickled and oiled (HRPO) and cold rolled, closed annealed (CRCA). It can be sheared on continuous rotary shears to the customer's specification where accurate control of the width is required. Alternatively, it can be left with a 'mill-edge' where the flat surfaces are bright rolled and the edges are left rounded and in the hot-rolled state. Non-ferrous strip is usually bright rolled and sheared to width. The rolling process tends to work harden the strip which is sold in various 'tempers' according to the amount of cold working it has received since annealing – for example, dead-soft, soft, quarter-hard, half-hard, etc.
- *Castings*   These can be made in most metals and alloys but the moulding process will vary depending upon the type of metal, the size of the casting, the accuracy of the casting and the quantities involved. There are no 'standard' castings; they are made to the customer's own patterns or, in the case of die casting, in the customer's own dies.
- *Tubes and pipes*   These can be made in ferrous and non-ferrous metals and alloys and in plastic. Tubes refer to the smaller sizes and pipes to the larger sizes. Steel pipes may be cold drawn or hot drawn without seams for the highest pressures. They may also be rolled from strip with a butt-welded seam running along the length of the pipe or tube where lower pressures are involved, or they may only be used for sheathing (electrical conduit is made like this). Plastic tube may be rigid or flexible.
- *Forgings*   These may be made by forming the red-hot metal with standard tools by a 'blacksmith' when only small quantities are required. The size may range from horse-shoes, farm implements, and decorative wrought-iron work made by hand, to turbine shafts and ships' propellor shafts forged under huge hydraulic presses. Where large quantities of forging of the same type are required, these are drop or press forged. The red-hot steel is forged in dies which impart the finished shape to the forging. Light alloy (aluminium alloy) forgings are used by the aircraft industry. Because of the low melting temperature of such alloys compared with steel, they have to be forged at a lower temperature. This requires a greater forging pressure and care must be taken to ensure that cracking does not occur.
- *Sheet and plate*   These start off as a very wide coiled strip which is passed through a series of rollers to flatten it. It is sheared to length by 'flying shears' that cut it whilst the flattened strip is moving. The terminology is somewhat vague and depends upon the metal thickness. Generally, 'sheet' can be worked with hand tools and ranges from foil (very thin sheet) to about 1.5 mm thick; 'thin plate' can be up to about 6 mm thick, after which it becomes thick plate. Both thin and thick plate have to be cut and formed using power-driven tools. Sheet is available in both cold- and hot-rolled finishes, whereas plate is only available as hot rolled. For the sizes of any standard metallic products, see BS 6722: 1986 (amended 1992).

6.12 Non-destructive testing

To check the suitability of materials used in engineering for a particular application, they can be subjected to:

- *Non-destructive testing*, where the component being tested is not broken, deformed or damaged in any way by the testing procedure. This is the type of testing to be considered in this section.
- *Destructive testing*, where the ability of a test specimen to withstand breaking and/or deforming forces are measured. These tests are described in Section 6.13.

A full programme of non-destructive testing can be very expensive, requiring each component to be handled individually through each of a series of tests. The tests themselves are often complex (radiography), requiring expensive equipment and highly skilled personnel to carry them out.

However, it is better to identify any defect at the start rather than discover a fault in the material after extensive and expensive machining operations have been completed.

For safety, welds on pipe runs and pressure vessels must be free from discontinuities, such as porosity and slag inclusions. Again, for safety, the forged aluminium alloy hinge components for aircraft control surfaces must be free from cracks due to forging at too low a temperature. Such defects can only be detected by non-destructive testing.

The non-destructive testing techniques to be considered in this section can be listed as:

- Visual examination.
- Sonic and ultrasonic testing.
- Magnetic testing.
- Radiography (X-rays).
- Macro-examination.

6.12.1 *Visual examination*

This is the cheapest and simplest possible form of non-destructive testing.

Straightness
Lack of straightness can be detected by sighting along a bar, as shown in Fig. 6.19(a) or, if the section is too heavy to lift, by use of a taut wire or cord, as shown in Fig 6.19(b). Since only the hollow side will show up in this test, check both sides of the component.

Scale
Excessive scale on the surface of a component can disguise surface cracks and other surface faults. It also damages the tips of cutting tools causing premature failure. Visual examination of castings can identify surface cracks, scabs, surface porosity, warping and twisting, and inadequate filling of the mould.

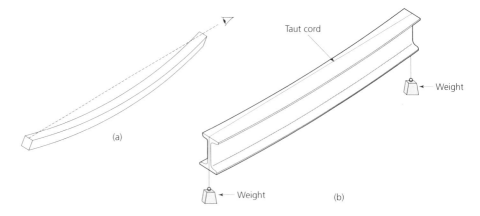

Welds

Visual inspection can show up such faults as surface cracks, inadequate fusion lack of penetration, insufficient filler material, poor joint shape, surface porosity and undercuts. Figure 6.20 shows the end views of two faulty welds whose faults can easily be identified by visual inspection.

Fig. 6.20 *Visual inspection of weld faults – a manual metal arc single 'V' butt weld showing: (a) a lack of penetration and misalignment of the plate edges; (b) undercut and misalignment of the plate edges*

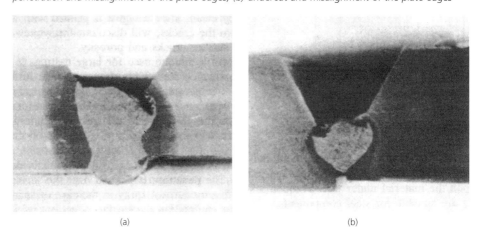

Dye penetrants

The coloured dye penetrant is sprayed onto the surface of the component from an aerosol can. The penetrant is absorbed into any surface cracks. Surplus penetrant is carefully wiped off the surface which is dried with a soft cloth. A developer is then sprayed onto the surface to leave a fine white powder as it dries. As the coloured penetrant oozes out of the crack it stains the white powder and shows clearly where the crack is. Alternatively, some penetrants glow when subjected to ultraviolet light. No developer is used. The penetrant is sprayed on and left for a short period to soak in. The surplus penetrant is wiped off, and an ultraviolet light source is shone onto the surface under inspection. Any cracks will be shown up by the glowing penetrant in the crack.

Safety
- Aerosols must only be pointed at the work, never at people.
- Carefully observe the manufacturer's instructions which must be read before use.
- If solvents are used to remove surplus penetrants, observe the manufacturer's instructions regarding ventilation. *Do not inhale the solvent.*
- If ultraviolet light is used, safety goggles with filter lenses must be worn and the company's safety regulations must be observed.

6.12.2 *Sonic and ultrasonic testing*

Sonic testing

A familiar sight at railway stations, some years ago, was the 'wheel tapper'. This was a railway official who walked along the train and tapped each wheel with a light hammer on a long handle. A cracked or defective wheel would give a dead note whilst a sound wheel would 'ring'. The defective item of rolling stock could then be taken out of service before an accident could occur.

Similarly, grinding wheels may be tested by 'ringing'. The wheel is suspended on a cord and lightly tapped with a wooden stick, as shown in Fig. 6.21. A sound wheel will 'ring' with a clear note, whilst a cracked wheel or a wheel containing a void will give a dead sound. It is only safe to fit a wheel that 'rings' with a clear sound on a grinding machine.

Fig. 6.21   *Checking a grinding wheel by 'ringing'*

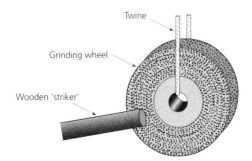

Twine

Grinding wheel

Wooden 'striker'

1. The wheel to be tested is freely suspended on stout twine.

2. It is **lightly** struck with a wooden rod.

3. If the wheel is free from cracks or manufacturing faults such as voids, it will 'ring' with a clear note.

Ultrasonic testing

Internal cracks in forgings and faulty welds together with blow holes in castings are much more difficult to detect. Ultrasonic inspection can detect defects inside materials. The highest frequency sound wave that most people can hear is around 18 kHz. The sound waves used for ultrasonic testing lie between 0.5 and 15 MHz, depending upon the material being tested. Frequencies between 1 and 3 MHz are usually used for steel.

The sound waves are generated electronically and transmitted as pulses of narrowly beamed sound waves. The higher the frequency, the more concentrated the beam of sound. The principle of operation is shown in Fig. 6.22. Normally only two pulses are shown on the screen of the test set: a pulse called the *initial incident pulse* from the 'front' surface component in contact with the probe, and a reflection from the 'rear' surface of the component. If there is a flaw in the component an additional pulse will appear on the screen. This will be the result of sound waves being reflected from the defect. The position and extent of the flaw can be determined by an experienced operator moving the probe. A set of equipment for ultrasonic testing is shown in Fig. 6.23.

Fig. 6.22 *Principles of ultrasonic testing: (a) paths of ultrasound waves; (b) VDU displays corresponding to the tests in (a)*

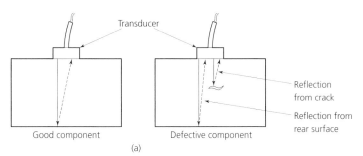

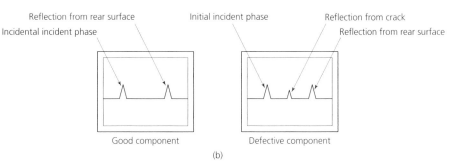

Fig. 6.23 *Single transducer set up for ultrasonic testing*

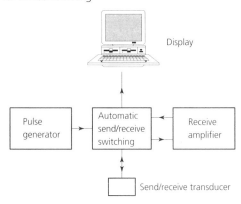

6.12.3 *Magnetic testing*

This is a simple and reliable means of detecting cracks and flaws at the surface and just below the surface of most steels and cast irons. It cannot be used with non-magnetic materials. Figure 6.24(a) shows the set up for this test. The component is placed between the poles of a powerful magnet. If there are no flaws in the component the magnetic flux field will flow straight through the component from pole to pole of the magnet. However, a flaw in the metal will cause the flux field to be distorted, as shown in Fig. 6.24(b). Finely spread iron powder will 'bunch' at these points indicating the presence of a crack or other flaw, as shown in Fig. 6.24(c).

For testing long bars of steel or cast iron (up to 4 metres long and 75 millimetres diameter) it is better to pass a heavy direct current at low voltage through the bar from end to end. The passage of the electric current through the bar generates a magnetic field around the bar. The bar is immersed in a shallow bath of iron powder in suspension in a liquid, and is slowly rotated in this suspension. Any defect will be shown up by bunching of the iron powder on the surface of the bar.

Fig. 6.24 *Principles of magnetic crack detection: (a) set up for magnetic crack detection; (b) distortion of the magnetic field caused by defects in the metal; (c) 'bunding' of the magnetic powder shows where the magnetic field is distorted*

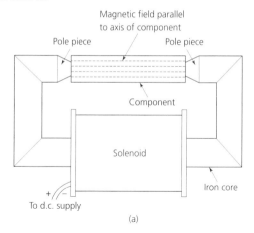

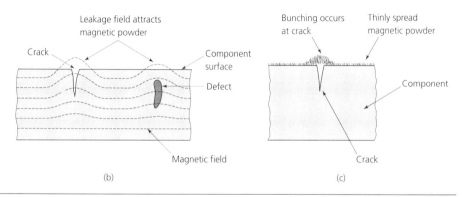

6.12.4 *Radiography (X-rays)*

This is a photographic process in which the 'illumination' of the component is by X-rays or by the even more penetrating gamma-rays. Any solid object between the radiation source and the photographic plate will absorb most or all of the X-rays and will cast a shadow on the plate. This shadow will be less dark than those areas of the plate that are exposed to the full strength of the X-rays, as shown in Fig. 6.25.

Fig. 6.25 *Principles of radiography: (a) general arrangement; (b) appearance of film after development*

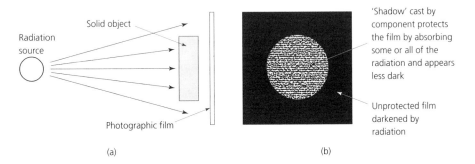

The thicker the material being inspected, or the more dense the material, the greater will be its absorption of the X-rays. Thus, if there is a void in the component, less absorption will take place at this point and the flaw will show up on the plate as a darker area, as shown in Fig. 6.26.

The smallest defect that can be readily detected by X-ray radiography is approximately 2 per cent of the section thickness. Below this the film or plate cannot resolve differences in radiation level – that is, for a component 25 mm thick, any flaw under 0.5 mm in size will remain undetected. Thus, the process becomes less reliable as the section thickness increases, and in this respect it is at a disadvantage compared with ultrasonic testing which can detect flaws in steel up to 150 mm thick with much less costly equipment. However, radiography has the advantage of providing a permanent visual record of the defect, its precise form and its precise location.

X-rays are generated electrically in high-voltage discharge tubes. This method of generation provides a constant source of radiation and the exposure time is easily controllable. Although widely used under laboratory or test department conditions, the size and weight of the equipment, and the need for an electrical supply, excludes its use for site work. Remember that the X-ray machines used for examining metal components are much more powerful than those used in hospitals for the examination of living tissue.

Gamma-rays are usually used for site work and thicker materials. They are much more powerful and penetrating than X-rays. The radiation source is a small amount of a radioactive substance such as iridium 192 or cobalt 60. These materials are constantly emitting radiation and must be stored in a heavy, lead-lined container. Provision is made for the opening of a 'window' by remote control in order to take the photograph. This is shown in Fig. 6.27.

Fig. 6.26 *Use of radiography to detect a casting defect: (a) general arrangement; (b) appearance of film after development*

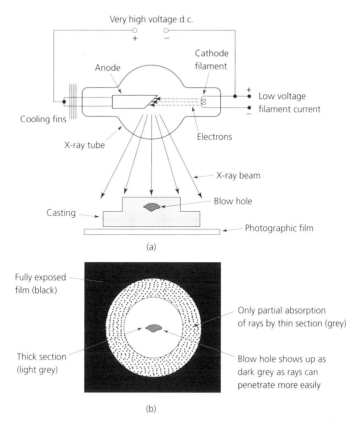

Fig. 6.27 *Typical gamma-ray source: (a) window closed – gamma-rays off; (b) window open – gamma-rays on*

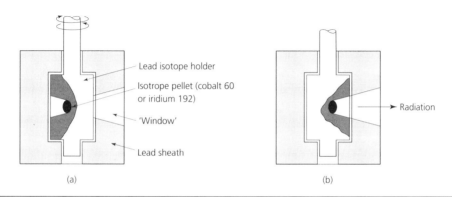

Stringent safety precautions and codes of practice must be observed wherever and whenever X-rays and gamma-rays are used. Operators of radiography equipment must be protected by radiation-resistant cover. Other persons must be kept well away from the radiation zone. Good ventilation is essential and the operator should change his or her overalls regularly. Normally the operator wears a radiation-sensitive tab on his or her overalls which changes colour if exposed to excess radiation, although this will be below the danger level for the operator. Radiographers should be given regular medical checks.

6.12.5 *Macro-examination*

The grain structure of metals and any inclusions and discontinuities which may be present can be studied by macro- or by microscopical examination. Only macro-examination is considered in this book.

Macro-examination implies the use of the unaided eye or the use of a low-power magnifying glass. The sample component is sectioned and the surface is ground smooth, taking care to use a free cutting wheel and coolant to avoid heating the surface of the material as heating may change the grain structure at the surface and give a false examination. Since grinding tends to 'drag' the surface, it is usual to hand finish the surface using a grade 0 or 00 abrasive paper. To reveal the grain structure it is necessary to *etch* the specimen. This is done by swabbing a suitable *etchant* onto the polished surface of the specimen. The etchant will eat away the material at its grain boundaries so that the individual grains stand out in relief. The specimen is then washed to stop the reaction. It is now ready for examination. Details can often be seen more clearly whilst the surface is still wet. Table 6.11 lists some suitable etchants for macro-examination. Figure 6.28 shows a typical example of the appearance of a component prepared for macro-examination. A slag inclusion in the weld is clearly visible and the specimen would be rejected. The general pattern of the grain structure of the weld metal is also clearly visible, as is the 'heat-affected' zone surrounding the weld metal.

Safety
Take care, etchants are corrosive:

- Do not get etchants on your clothing – wear a plastic apron.
- Do not get etchants on your skin – wear disposable plastic gloves.
- In the event of a splash, wash the affected area under running cold water and seek medical attention.
- Do not get etchants in your eyes – wear goggles.
- In the event of a splash, wash your eye with running cold water or an eye irrigation bottle and seek immediate medical attention.

Table 6.11 *Etchants for macro-examination*

Material	Composition	Application
Steel	50% hydrochloric acid (conc.) 50% water	Specimen boiled in etchant for 5-15 minutes. For revealing flow lines, structure of fusion welds, cracks, porosity, case depth
	25% nitric acid (conc.) 75% water	As above, but can be applied by cold swabbing for large specimens
	Stead's reagent	Reveals dendritic structure in steel castings, and phosphorous segregation
Aluminium and aluminium alloys	20% hydrofluoric acid (conc.) 80% water	Reveals flow lines and general grain structure and impurities (undissolved inclusions)
	45% hydrochloric acid (conc.) 15% hydrofluoric acid (conc.) 15% nitric acid (conc.) 25% water	As above, but more reactive reagent. Avoid contact with skin
Copper and copper alloys	25 g ferric chloride in 100 ml 25% hydrochloric acid (conc.) 75% water	Reveals dendritic structure of α-phase solid solutions
	33% ammonium hydroxide (0.880) 33% ammonium persulphate (5%) 34% water	Reveals β-phase structure

Fig. 6.28 *Specimen as it appears for macro-examination*

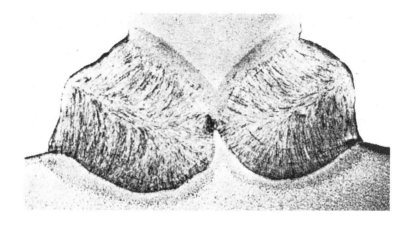

6.13 Destructive testing

Destructive testing is performed on prepared specimens that differ according to the test to be carried out. These tests are used to determine and measure the properties of a given material. The tests to be considered in this section are:

- Tensile testing.
- Bend testing.
- Impact testing (Izod and Charpy techniques).
- Hardness testing (Brinell, Vickers and Rockwell techniques).

6.13.1 *Tensile testing*

Strength is defined as the ability of a material to resist applied forces without yielding or fracturing. Strictly speaking, this definition applies equally well to strength and toughness. however, by convention, *strength* usually refers to the resistance of a material to withstand a tensile (pulling) load – see Section 6.2. *Toughness* refers to the resistance of a material to impact loads and will be dealt with later. Figure 6.29 shows a typical tensile specimen and the basic technique for carrying out a tensile test.

Fig. 6.29 *Tensile test: (a) typical specimen; (b) principle of tensile test*

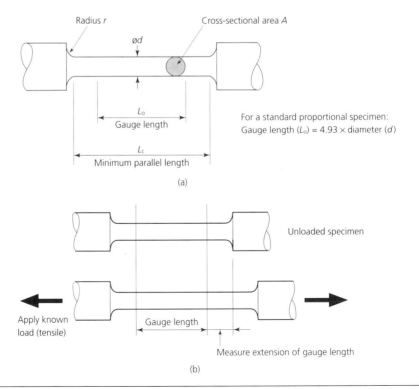

The specimen will have been manufactured to the appropriate British Standard. When the load is applied, the specimen starts to stretch. The amount by which it stretches is measured between the points referred to as 'gauge length' in Fig. 6.29. If the material has *elastic* properties, the specimen will return to its original length when the load is removed. If the material has *ductile* properties it will *not* return to its original length when the load is removed. It will remain at the length to which it has been stretched. To perform the test, the tensile load on the specimen gradually increases and the extension of the specimen is measured for each increase in load. This continues until the specimen breaks. Figure 6.30 shows two typical testing machines. Both these machines steadily increase the load on the specimen at a prescribed rate and, at the same time, automatically plot a graph of extension against load.

Fig. 6.30 *Tensile testing machines: (a) the TecQuipment (TASK) MF2 Mini Tensile Tester (source: TecQuipment Ltd); (b) the Universal testing machine (source: Samuel Denison Ltd)*

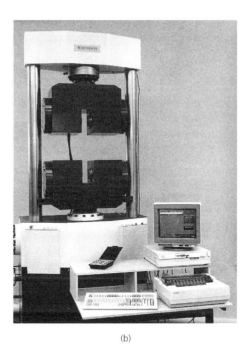

(a)   (b)

Let's assume we are testing a specimen made from an annealed sample of a low-carbon steel. The graph will be similar to the one shown in Fig. 6.31. I have shown the graph plotted using *load* and *extension*.

• From A to B the graph is a straight line. This indicates that the extension is proportional to load and the material is behaving in an *elastic* manner. If the load is removed the specimen returns to its original length.

Fig. 6.31 *Load–extension curve for low-carbon steel*

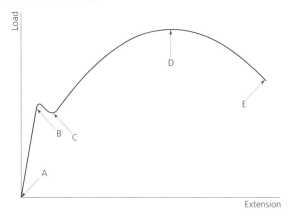

- At B the graph ceases to be a straight line; we have reached the *limit of proportionality*.
- From B to C we get a 'kink' in the graph. The specimen suddenly appears to give way. This is called the *yield point*. The specimen will no longer spring back to its original length. It has taken a *permanent set*.
- From C to D the extension is no longer proportional to load and, if the load is removed, little or no spring back will occur. The specimen is now showing *plastic* properties. It is behaving in a *ductile* manner.
- The point D is referred to as the *ultimate tensile stress* (UTS) for the material. Although useful for comparing material properties, it has little practical value since engineering equipment is not usually operated at its breaking point.
- From D to E the specimen appears to continue stretching under a reduced load. In fact, the specimen is thinning out, as shown in Fig. 6.32(a). This is called 'necking'. This localised thinning results in the 'load per unit area' or *stress* increasing despite the fact that the *load* is getting less. The specimen finally breaks (fractures) at the point E in Fig. 6.31, as shown in Fig. 6.32(b).

Figure 6.32 shows the load–extension curve for an annealed mild steel specimen. In fact values of load and extension are of little use to us since they only apply to a specimen of one particular size. We need to convert load to stress and extension to strain as follows:

$$\text{stress} = \frac{\text{applied load}}{\text{the original cross-sectional area of the specimen}}$$

$$\text{strain} = \frac{\text{extension}}{[\text{the original gauge length of the specimen}}$$

Figure 6.33 shows some typical curves for a range of materials. This time the axes of the graphs have been labelled *stress* and *strain*. You are already familiar with the curve for mild steel. Compared with the steel curve, the cast-iron curve shows little extension, virtually no plastic deformation and a low stress value at the point of fracture. It also has a very 'upright' initial straight section typical of rigid material. The light alloy curve shows an initial straight section that is far less upright, indicating a material lacking in rigidity. It has an extended plastic zone typical of a light alloy that is soft and ductile.

Fig. 6.32 *Necking and fracture of a cylindrical specimen*

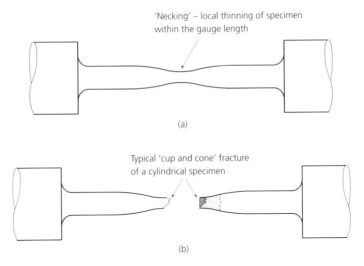

'Necking' – local thinning of specimen within the gauge length

(a)

Typical 'cup and cone' fracture of a cylindrical specimen

(b)

Fig. 6.33 *Typical stress–strain curves: (a) for annealed mild steel; (b) for grey cast iron; (c) for a light alloy*

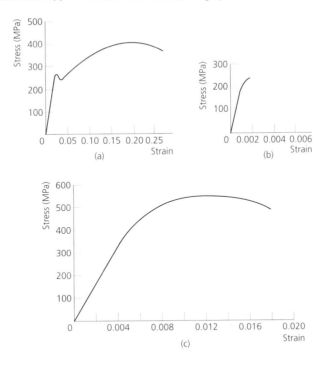

Factor of safety

Usually a designer proportions a component so that the maximum working *stress* does not exceed 50 per cent of the *stress* at the limit of proportionality for the material.

Ductility

In addition to determining the tensile strength of the specimen, the tensile test may also be used to determine the *ductility* of the material. The ductility of a material is proportional to the *percentage elongation* of the fractured test piece.

$$\text{elongation } \% = \frac{\text{increase in length}}{\text{original length}} \times 100$$

The increase in length is determined by fitting the pieces of the fractured specimen together again and measuring the length at failure.

$$\text{increase in length } = \text{ length at failure} - \text{original length}$$

Proof stress

Only very ductile materials, such as annealed low-carbon steel, show a marked yield point. It may not even appear for a bright-drawn steel because of the effect of work hardening. Under such conditions the *proof stress* is used. Figure 6.34 shows a typical stress–strain curve for a material of relatively low ductility such as a hardened and tempered medium-carbon steel. The line BC has been drawn parallel to the straight portion of the curve from the point B. This point is fixed by the specified conditions for the test. For example, if 0.2 per cent proof stress is required, AB represents the strain resulting from an extension of 0.2 per cent of the gauge length of the specimen. The corresponding stress at C will be *the 0.2 per cent proof stress*. The material will have met its specification if, after the proof stress has been applied for 15 seconds and then removed, the permanent set of the specimen has not exceeded 0.2 per cent of the gauge length.

Fig. 6.34 *Proof stress*

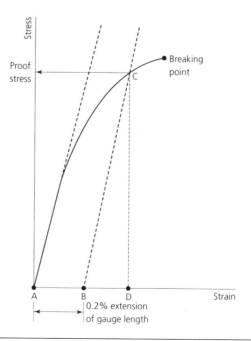

6.13.2 *Bend testing*

The percentage elongation of a material, as determined by the tensile test, has already been discussed as a measure of ductility. The ductility of materials can also be compared and assessed by simple workshop bend tests. There are several ways in which the tests can be applied and these are shown in Fig. 6.35. The test chosen will depend upon the ductility of the material and the severity of the test applied.

Fig. 6.35 *Bend test: (a) close bend test; (b) angle bend test; (c) 180° bend test*

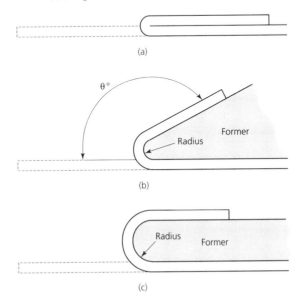

- *Close bend test*   This is shown in Fig. 6.35(a). The specimen is bent over upon itself and flattened. No allowance is made for spring back, and the material is satisfactory if the test can be completed without the material tearing or fracturing.
- *Angle bend test*   This time the material is bent over a former, as shown in Fig. 6.35(b). The nose radius of the former and the angle of bend (h) are fixed by specification. No allowance is made for spring back and the material is considered to be satisfactory if the test can be completed without the material tearing or fracturing.
- *180° bend test*   This is a development of the angle bend test using a flat former and only the nose radius is specified, as shown in Fig. 6.35(c). No allowance is made for spring back and the material is considered satisfactory if the test can be completed without the material tearing or fracturing.
- *Reverse bend test*   This is used for very ductile materials where the preceding tests are not severe enough. The material is bent repeatedly, back and forth, round a former of specified nose radius through 90° or 180° for a specified number of times.

In all the above tests the test piece is a strip of the material 10 mm wide, with rounded edges.

6.13.3 *Impact testing*

The tensile test alone does not tell the whole story. Figure 6.36(a) shows how a piece of high-carbon steel in the annealed condition bends when struck with a hammer. Figure 6.36(b) shows how a similar piece of high-carbon steel that has been hardened by heating and quenching, will more easily fracture when struck by a hammer because it is now brittle. Yet in the hardened and relatively brittle condition this piece of high-carbon steel will have a higher tensile strength than when it was in the annealed and more malleable condition. It has gained in 'strength' but has lost its 'toughness'.

Fig. 6.36 *Impact loading: (a) a piece of high-carbon steel rod (0.1%) in the annealed (soft) condition will bend when struck with a hammer – UTS 925 MPa; (b) the same piece of high-carbon steel rod after hardening and lightly tempering will fracture when hit with a hammer, despite its UTS having increased to 1285 MPa*

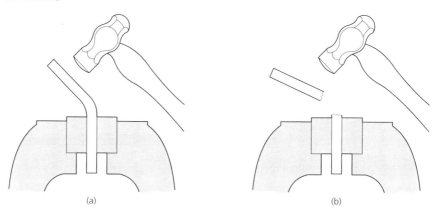

(a)　　　　　　　　　　　(b)

Impact testing consists of striking a suitable specimen with a controlled blow and measuring the energy absorbed in bending or breaking the material. This energy value indicates the *toughness* of the material – that is, its ability to withstand transverse impact loading. Figure 6.37 shows a typical impact testing machine.

- *Izod test*　In this test a 10 mm square, notched specimen is used. The specimen is held in a vice, as shown in Fig. 6.38(a). The striker hits the specimen with a kinetic energy of 162.72 joules (J) at a velocity of 3.8 metres per second (m/s).
- *Charpy test*　This is now the more widely used test. In the Izod test the specimen was supported as a *cantilever* (held only at one end). In the Charpy test it is supported as a *beam* – that is, it is held at both ends. It is struck with a kinetic energy of 298.3 joules (J) at a velocity of 5 metres per second (m/s). Figure 6.38(b) shows details of the specimen and the manner in which it is supported.

In both tests a notched specimen is used. Therefore useful information can be obtained regarding the resistance of a material to the spread of a crack. The crack usually originates from a point of stress concentration and indicates the need to avoid sharp corners, notches, undercuts, sudden changes in section and machining marks in heavily stressed components.

Fig. 6.37 *Typical impact testing machine (source: Samuel Denison Ltd)*

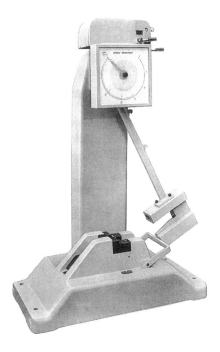

Safety

When these tests are carried out, particularly on brittle materials, broken pieces of material may fly off with considerable force. The test zone should be guarded and protective goggles or a visor should be worn.

6.13.4 *Hardness testing*

Hardness has already been defined as the resistance of a material to indentation by another hard body. This is precisely how hardness is measured.

Brinell hardness test

In this test, a hardened steel ball of specified size is pressed into the surface of the material under test by a specified load, as shown in Fig. 6.39(a). The relationship between the load (P, kg) and the diameter (D, mm) of the ball indenter is given by the expression:

$$\frac{P}{D^2} = K$$

where K is a constant. Typical values of K for various materials are:

Ferrous metals	$K = 30$
Copper and copper alloys	$K = 10$
Aluminium and aluminium alloys	$K = 5$
Lead, tin and white-bearing metals	$K = 1$

Fig. 6.38 *Impact test conditions: (a) Izod test; (b) Charpy test. (Dimensions in millimetres)*

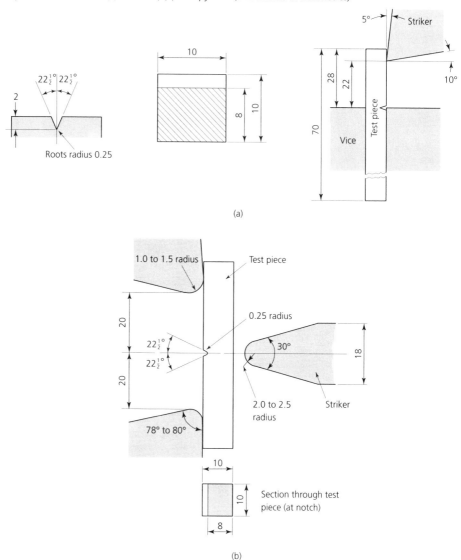

(a)

(b)

Since the calculation of the spherical area of the indentation is not very easy, the hardness number is obtained as follows:

- The diameter of the indentation is measured by means of a low-powered microscope fitted with a scale. The appearance of the indentation and the scale as it appears through the microscope is shown in Fig. 6.39(b). Take the average of two readings at right angles to each other.
- The Brinell hardness number (H_B) is obtained by the use of conversion tables that convert the average diameter of the indentation into the hardness number without

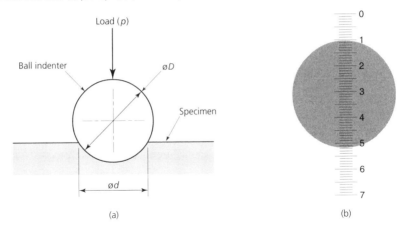

(a) (b)

calculation. Separate tables are provided for each standard combination of load and size of indenter.

Note: It is necessary to take certain precautions to obtain an accurate hardness reading. These precautions are as follows.

- The thickness of the specimen should be at least seven times the depth of the indentation.
- The edge of the indentation should be at least three times the diameter of the indenter from the nearest edge.
- This test is unsuitable for materials whose hardness exceeds 500 H_B, as the ball will tend to flatten and give a false result.

Machinability
With high-speed steel-cutting tools, the hardness of the material being cut should not exceed $H_B = 350$ for a reasonable tool life. Again, materials with a hardness less than $H_B = 100$ will tend to tear and leave a poor finish.

Work-hardening capacity
Materials that will cold work without work hardening unduly will pile up around the indenter, as shown in Fig. 6.40(a). Materials that will work harden readily will sink around the indenter, as shown in Fig. 6.40(b).

Fig. 6.40 *Work hardening capacity: (a) piling up; (b) sinking*

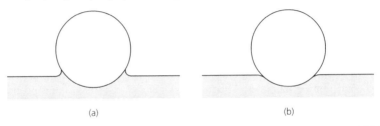

(a) (b)

Vickers hardness test

This test is preferable to the Brinell test where hard materials, such as cutting tool materials, are being tested. This is because a diamond is used for the indenter in this test. The diamond is in the form of a square-based pyramid with an angle of 136° between any pair of opposite faces. Since only one size of indenter is used for this test, the load has to be varied for different hardness ranges for different material. Standard loads are: 5, 10, 20, 30, 100 kg. For example $H_D(50) = x$ indicates that a 50 kg load has been used. The hardness number is x. Figure 6.41(a) shows a typical Vickers hardness-testing machine and Fig. 6.41(b) shows the appearance of the indentation. The indentation is measured across its corners. Two readings are taken at right angles to each other and the average reading is used. The hardness number is obtained from conversion tables.

Fig. 6.41 *Vickers hardness test: (a) testing machine; (b) measuring screen showing magnified image of impression (source: Samuel Denison Ltd)*

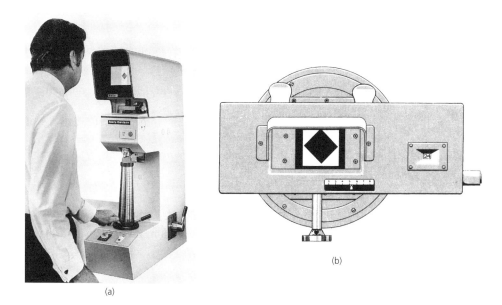

(b)

(a)

Rockwell hardness test

Although not as accurate as the Brinell or Vickers tests for laboratory purposes, the Rockwell test is widely used in industry, in heat-treatment shops, as it is quick and easy to use. This test registers the hardness reading directly on a dial without the need for conversion tables or calculations. A steel ball indenter is used for soft materials, and a diamond cone (120°) indenter is used for hard materials. The test consists of applying a minor load and setting the dial to zero. The major test load is then added and the depth of indentation increases. The hardness number registered is proportional to the increase in penetration of the indenter. Figure 6.42(a) shows a typical Rockwell machine and

Fig. 6.42(b) shows the appearance of the hardness indicating dial. Test conditions for most common scales of hardness are:

- 'B' scale for soft materials: 1.6 mm diameter ball
  100 kg major load
- 'C' scale for hard materials: 120° diamond cone
  150 kg major load

The minor load in both cases is 10 kg.

As for all branches of engineering, the tests described in this section are becoming 'computerised'. The latest test machines perform the test and compute the test results automatically without the need for conversion tables. Hard copy printouts of the results are usually available and graphs of stress and strain will be produced automatically in the case of tensile tests. However, many manual machines are still in existence, and will be for many years to come, which is why they are included in this chapter.

Fig. 6.42 *Rockwell hardness test: (a) testing machine; (b) direct reading hardness scale (source: Samuel Denison Ltd)*

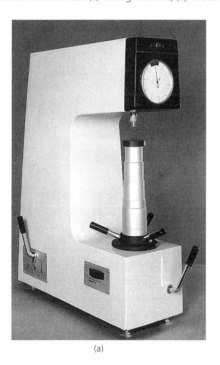

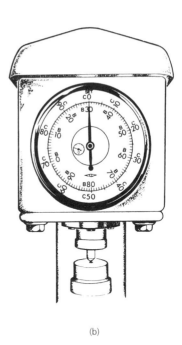

(a)   (b)

Most hardness-testing machine manufacturers publish tables showing comparative hardness values for various methods of testing. Such comparative tables should be treated with caution since the tests are carried out under different conditions. Figure 6.43 shows the approximate relationship between typical hardness scales.

Fig. 6.43 *Hardness scales*

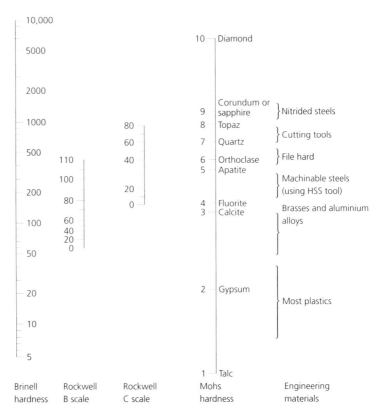

6.14 Identification of ferrous metals

6.14.1 *Colour codes*

The similarity between many materials of different physical properties makes it essential that some sort of permanent identification should be marked on them, e.g. colour coding. Usually the colour is painted on bar ends and as a triangle on the corners of sheets. It is then easy to identify the material providing you know the code. Unfortunately, although many firms use colour coding in their stores, there is no standard system of coding. You have to learn the code used in your particular place of work. The coding may even differ between different branches of the same firm. There are exceptions: for example, British Standard 1490: 1988 specifies colour codes for aluminium and aluminium ingots and castings. However, bar 'ends' are often used up for 'one-off' jobs and Table 6.8 in Section 6.8 listed some simple workshop tests for material identification Such tests are not foolproof and require some experience.

6.14.2 *Abbreviations*

Table 6.12 lists some of the abbreviations used for ferrous metals. They may be found on storage racks in the stores and on engineering drawings. With the exception of 'silver steel', such abbreviations are very imprecise and refer mainly to groups of materials that can vary widely in composition and properties within the group. It is better to specify a material precisely using a British Standard coding.

Table 6.12 *Abbreviations for ferrous metals*

Abbreviation	Metal
CI	Cast iron (usually 'grey' cast iron)
SGCI	Spheroidal graphite cast iron
MS	Mild steel (low-carbon steel)
BDMS	Bright drawn mild steel
HRPO	Hot-rolled pickled and oiled mild steel
CRCA	Cold-rolled close annealed mild steel
GFS	Ground flat stock (gauge plate)
LCS	Low-carbon steel
SS	Silver steel (centreless ground high-carbon steel)
HSS	High-speed steel
Bright bar	Same as BDMS
Black bar	Hot-rolled steel still coated with scale

6.14.3 *British Standards for wrought steels*

During the Second World War all wrought steels were standardised in BS 970, and the steels were given EN numbers. The initials EN stood for either *emergency number* or for *economy number,* the exact meaning having become lost in the mists of time. This was a random system of numbering. EN8 being a medium carbon steel, EN32 a case-hardening low-carbon steel and EN24 a high-tensile nickel–chrome alloy steel. Despite the fact that the BSI issued memoranda to all major suppliers and users of wrought steel stating that the old system should be discontinued as soon as possible after 1972 to avoid confusion, these out-dated specifications are still widely used.

Between 1970 and 1972, BS 970 was reissued using number and letter codes that more accurately described the composition and properties of the steels listed. This code is built up as follows.

The first three symbols are a number code indicating the type of steel.

- 000 to 199   Carbon and carbon manganese steels. The numbers indicate the manganese content ×100.
- 200 to 240   Free-cutting steels, with the second and third numbers indicating the sulphur content ×100.
- 250   Silicon valve steels.

- 300 to 499   Stainless steels.
- 500 to 999   Alloy steels.

The fourth symbol is a letter code and is applied as follows:

- A   The steel is supplied to a chemical composition determined by chemical analysis of a batch sample.
- H   The steel is supplied to a hardenability specification. This is the maximum hardness that can be obtained for a specimen of a specified diameter.
- M   The steel is supplied to a mechanical property specification.
- S   The material is a stainless steel.

The fifth and sixth symbols are a number code indicating the average carbon content for a given steel The code is carbon content $\times 100$.

Therefore a steel that is specified as BS970.040A10 is interpreted as:

- BS970   indicates the standard being applied.
- 040   lies between 000 and 199 and indicates that we are dealing with a plain carbon steel containing some manganese.
- 040   also indicates that the steel contains 0.40 per cent manganese since $0.40 \times 100 = 040$.
- A   indicates that the composition has been determined by batch analysis.
- 10   indicates that the steel contains 0.1 per cent carbon since $0.1 \times 100 = 10$.

BS970 is currently issued in four parts:

- Part 1:   *General inspection and testing procedures and specific requirements for carbon, carbon manganese, alloy and stainless steels.*
- Part 2:   *Requirements for steels for manufacture of hot formed springs.*
- Part 3:   *Bright bars for general engineering purposes.*
- Part 4:   *Valve steels.*

BS970 is slowly being phased out and replaced by a new family of standards covering all aspects of steel making and supply. These range between BS EN 10001 and BS EN 10237 inclusive. The initials EN now indicate that these standards are the English language versions of European Standards (EN = European number). These new standards are outside the scope of this book.

EXERCISES

6.1   *Material properties*
  (a) Name the properties required by the materials used in the following applications:
  (i) a metal cutting tool
  (ii) a forged crane hook
  (iii) a motor car radiator
  (iv) a motor car road wheel axle

(v) the conductors in an electric cable
(vi) a crane sling
(vii) the sheathing of an electric cable
(viii) a kitchen sink
(ix) a garden hose pipe
(x) concrete for a machine foundation

(b) Figure 6.44 shows a number of material applications as listed below. Name suitable materials for each of the applications, giving reasons for your choice:
(i) conductors of an electric cable
(ii) insulation and sheathing of an electric cable
(iii) moulding for the connector block
(iv) screws and inserts for the connector block
(v) connecting rod for a car engine
(vi) lathe tailstock

Fig. 6.44 *Exercise 6.1(b)*

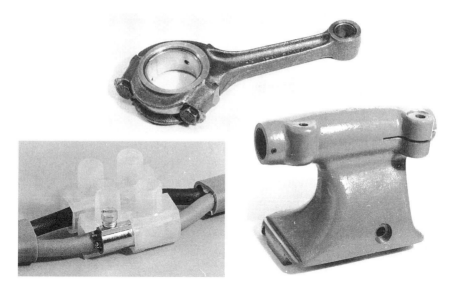

(c) Giving reasons for your choice, name a suitable plain carbon steel and state its heat-treatment condition for each of the following applications:
(i) cold chisel
(ii) engineer's file
(iii) vehicle leaf spring
(iv) sheet steel for pressing out car body panels
(v) rod for making small turned parts on an automatic lathe

6.2 *Material applications and classification*

    (a) Copy out and complete Table 6.13.

    (b) Copy out and complete Table 6.14 by explaining briefly the meaning of the terms

Table 6.13 *Exercise 6.2(a)*

Material	Typical application
Cast iron	
High-speed steel	
Duralumin	
Stainless steel (austenitic)	
Gunmetal	
Phosphor bronze	
70/30 Brass	
60/40 Brass	
Free-cutting brass	
Tufnol	
Nylon	
PTFE	
Perspex	
Polystyrene	
PVC	
Grass fibre-reinforced polyester	
Epoxy resin	
Urea–formaldehyde	

Table 6.14 *Exercise 6.2(b)*

Term	Meaning	Example
Ferrous metal		
Non-ferrous metal		
Thermoplastic		
Thermosetting plastic		
Synthetic material		
Natural material		
Metallic		
Non-metallic		
Alloy		

6.3   *Forms of supply, identification and specification*

   (a)   Table 6.15 lists a number of material applications. Copy out and complete the table by naming the 'form of supply' in which you would expect to receive the material for each application.

Table 6.15 *Exercise 6.3(a)*

Applications	Form of supply
Car body panels	
Lathe bed	
Turned parts	
Plastic mouldings	
Structural steel work	
Electric cables	
Printed circuit boards	
The two main raw materials for GRP boat hull mouldings	
Plastic window frames	
Connecting rods for high-power engines	

   (b)   (i)   State the meaning of the following abbreviations as applied to plain carbon steels: MS, BDMS, HRPO, CRCA.

         (ii)   What do the terms 'quarter-hard', 'half-hard', etc., refer to when ordering non-ferrous sheet metal and rolled strip?

   (c)   Describe the methods of material identification used in the raw material stores at your place of work or your training workshop.

   (d)   The following specifications are based upon the current edition of BS970. Explain their meaning:

         (i)   080A15
         (ii)   080M15
         (iii) 230M07
         (iv)   230M07 (leaded)
         (v)   080M40
         (vi)   605M36
         (vii) 708M40
         (viii) 817M40

6.4   *Checking and testing materials*

   (a)   Explain briefly the main differences between 'destructive' and 'non-destructive' testing and give an example where each would be used.

(b) Describe how you could:
    (i) check the straightness of a steel rod 10 mm diameter by 2 m long
    (ii) check a casting for blow holes just below the surface
    (iii) check a grinding wheel for cracks before mounting it on a grinding machine spindle
(c) Figure 6.45 shows the graph obtained by a destructive test on a sample of a material:
    (i) name the test
    (ii) explain the significance of the straight line part of the graph between the points A and B
    (iii) name the point C
    (iv) state what has happened at the point D
    (v) describe how this test can also be used to give an indication of the ductility of the metal being tested

Fig. 6.45 *Exercise 6.4(c)*

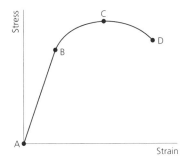

(d) With the aid of sketches, describe briefly **one** of the following tests, stating the property being tested:
    (i) Izod test
    (ii) Charpy test
(e) With the aid of sketches, describe a hardness test with which you are familiar.

7 Heat treatment

When you have read this chapter, you should understand:

- Safe working practices as applicable to heat-treatment processes.
- The principles and purposes of heat treatment.
- The through hardening of plain carbon steels.
- The flame hardening of medium/high-carbon steels.
- The carburising and case hardening of low-carbon steels.
- How to temper hardened steels.
- How to anneal and normalise steels.
- The basic heat treatment of non-ferrous metals and alloys.
- The principles, advantages and limitations of heat-treatment furnaces.
- Heat-treatment defects and the importance of furnace atmosphere control.
- The temperature control of heat-treatment furnaces.
- The advantages, limitations and applications of quenching media.

7.1 Introduction

Some heat-treatment processes, as a means of modifying the properties of metals, have already been introduced in Chapter 6. As a reminder, Table 7.1 summarises the more common heat-treatment processes for plain carbon steels. These and other processes will now be considered in detail in this chapter.

Because of the wide range of non-ferrous metals and alloys that exist, the heat-treatment processes for non-ferrous metals vary widely and all such processes are quite different to the processes used for the heat treatment of plain carbon steels. Some of the more important processes for the heat treatment of copper-based and aluminium-based alloys will also be included in this chapter.

Table 7.1 *Heat-treatment definitions*

Term	Meaning
Annealed	The condition of a metal that has been heated above a specified temperature, depending upon its composition, and then cooled down in the furnace itself or by burying it in ashes or lime. This annealing process makes the metal very soft and ductile. Annealing usually precedes flow-forming operations such as sheet metal pressing and wire and tube drawing
Normalised	The condition of a metal that has been heated above a specified temperature, depending upon its composition, and then cooled down in free air. Although the cooling is slow, it is not as slow as for annealing so the metal is less soft and ductile. This condition is not suitable for flow forming but more suitable for machining. Normalising is often used to stress relieve castings and forgings after rough machining
Quench hardened	The condition of a metal that has been heated above a specified temperature, depending upon its composition, and then cooled down very rapidly by immersing it in cold water or cold oil. Rapid cooling is called **quenching** and the water or oil is called the **quenching bath**. This rapid cooling from elevated temperatures makes the metal very hard. Only medium- and high-carbon steels can be hardened in this way
Tempered	Quench-hardened steels are brittle as well as hard. To make them suitable for cutting tools they have to be reheated to a specified temperature between 200 and 300 °C and again quenched. This makes them slightly less hard but very much tougher. Metals in this condition are said to be *tempered*

7.2 Heat-treatment process safety

General safety, including fire prevention and fire fighting, is included in Chapter 8. However, heat treatment can involve large pieces of metal at high temperatures and powerful furnaces. Therefore it is necessary to consider some safety practices relating specifically to heat-treatment processes.

7.2.1 *Protective clothing*

Ordinary workshop overalls do not offer sufficient protection alone. Many are made from flammable cloths. Further, synthetic cloths made from nylon or rayon fibres, or mixtures of natural and synthetic fibres, can melt and stick to your skin at the temperatures met with in heat treatment. This worsens any burn you may receive. Overalls used in heat-treatment shops should be made from a flame-resistant or a flame-retardant material and be labelled accordingly. In addition, a leather apron should be worn to prevent your overalls coming into contact with hot workpieces and hot equipment. An example is shown in Fig. 7.1.

Fig. 7.1 *Protective clothing (heat treatment)*

Safety helmet or cap made from flame retardant material

Overalls made from flame retardant material

Leather apron

Industrial safety shoes or boots

7.2.2 *Gloves*

Gloves should be worn to protect your hands. These should be made from leather or other heat-resistant materials and should have gauntlets to protect your wrists and the ends of the sleeves of your overalls. Leather gloves offer protection up to 350 °C. Suitable gloves are shown in Fig. 7.2.

Fig. 7.2 *Gloves suitable for heat treatment: (a) leather glove with reinforced palm – ideal for handling steel and sections; (b) gauntlet – available in rubber, neoprene or PVC for handling chemical, corrosive or oily materials; (c) heat-resistant leather glove – can be used for handling objects heated up to 360 °C; (d) chrome leather hand-pad or palm – very useful for handling sheet metal or glass; (e) industrial gauntlet – usually made of leather because of its resistance to heat; gauntlets not only protect the hands but also the wrists and forearms from splashes of molten salts and hot-quenching media.*

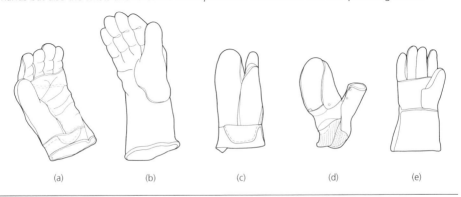

(a)   (b)   (c)   (d)   (e)

7.2.3 *Visors, goggles and headwear*

Visors, goggles and headwear should be worn when there is any chance of danger to your eyes, the skin of your face and your hair and scalp. Such dangers can come from various sources:

- Splashes from the molten salts when using salt-bath furnaces.
- Splashes from hot liquids when quenching.
- The accidental ignition of oil-quenching baths due to overheating.
- Radiated heat from large furnaces when their doors are opened.
- Accidental 'flash-backs' when lighting up furnaces.

Headwear should be in the form of a cap made from a flame-resistant or flame-retardant material. Better, are the industrial safety helmets of the type shown in Fig. 7.3(a). These helmets are described in greater detail in Chapter 8. Some examples of visors and goggles are shown in Fig. 7.3(b) and (c).

Fig. 7.3 *Head and eye protection: (a) a typical fibre-glass safety helmet made to BS 2826; (b) plastic face safety visor for complete protection against chemical and salt-bath splashes; (c) transparent plastic goggles suitable for machining operations*

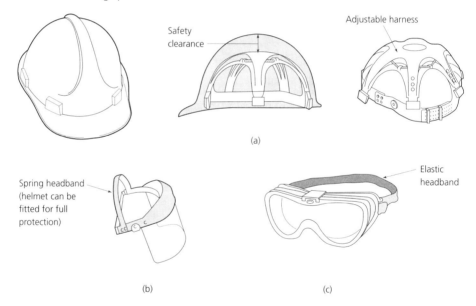

7.2.4 *Safety shoes and boots*

The safety shoes and boots recommended for workshop wear in Chapter 8 are also most suitable for use in heat-treatment shops. They not only protect you from cuts and crushing from heavy falling objects, being made of strong leather they also protect against burns from hot objects. Remember, your first instinctive reaction to picking up anything hot accidentally, is to let it go quickly and drop it. This is when the toe protection of industrial safety shoes earn their keep. Examples of such footwear are shown in Chapter 8. In addition, it is advisable that leather spats are worn. These protect your lower legs and ankles from splashes of molten salts or spillage of hot-quenching fluids. Spats are particularly important if you wear safety shoes rather than safety boots.

7.2.5 *Safety equipment*

Heat treatment can cover a whole range of processes and sizes of workpieces, from hardening a simple tool made from a piece of silver steel, using a gas blowpipe to heat it and a bucket of cold water to quench it, to the production treatment of large components. They all have the same basic problem – that is, very hot metal has to be handled. There are very many methods of handling hot workpieces, depending upon their size, the quantity involved, type of furnace being used and the process. Here are a few examples of how the hot components can be handled:

- Small components can be handled individually using tongs, as shown in Fig. 7.4(a).
- Quantities of small components can be handled on trays or in baskets. These trays and baskets are usually made from a heat-resistant metal such as the nickel alloy called *inconel*. The baskets can be handled in or out of salt bath furnaces and in or out of the quenching bath using a hoist.
- Large components and trays of smaller components can be handled in or out of muffle type furnaces on roller conveyors, as shown in Fig. 7.4(b). Long-handled hooks and rakes are used to push the components into the furnace or to pull them out when hot.

Fig. 7.4 *Handling hot components: (a) tongs for holding small components; (b) roller conveyor*

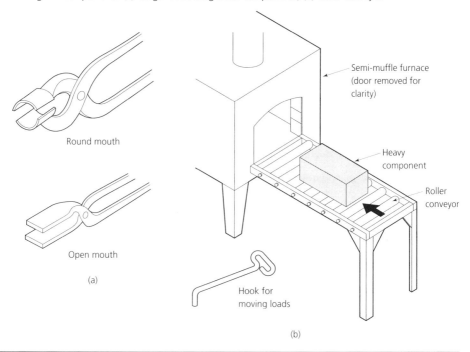

Round mouth

Open mouth

(a)

Semi-muffle furnace (door removed for clarity)

Heavy component

Roller conveyor

Hook for moving loads

(b)

7.2.6 *Safety notices*

Whilst metal is red hot it is obviously in a dangerous condition. However, most accidents occur when the metal has cooled down to just below red heat. Although no longer glowing, it is still hot enough to cause serious burns and to start fires if flammable substances come into contact with it. Hot workpieces must never be stored in gangways and warning notices must be used, as shown in Fig. 7.5. Such notices must satisfy the legal requirements of the Health and Safety Executive.

Fig. 7.5 *Safety notices must be placed by hot objects*

7.2.7 *Fire*

Quenching baths using a quenching oil must have an airtight lid, as shown in Fig. 7.6. In the event of the oil overheating and igniting, the lid can be closed and the lack of air will extinguish the fire. Quenching tanks should always have sufficient reserve capacity so that they do not overheat. Large tanks being used on production work must be kept topped up and a pyrometer (high-temperature thermometer) should be fitted so that the temperature of the quenching oil can be checked. If it gets too hot:

- The oil will not cool the work quickly enough to harden it.
- The oil may catch fire.

Only quenching oil with a high flash point and freedom from fuming should be used. *Lubricating oil must never be used.* A suitable fire extinguisher, or several extinguishers if a large quenching tank is used, should be positioned conveniently near to the bath for use in an emergency. The type of extinguisher should be suitable for use on oil fires. Fire extinguishers are dealt with in Chapter 8.

Furnaces and blowpipes must not be lit or closed down without proper instruction and permission. Incorrect setting of the controls and incorrect lighting up procedures can lead to serious explosions.

Fig. 7.6 *Quenching tank*

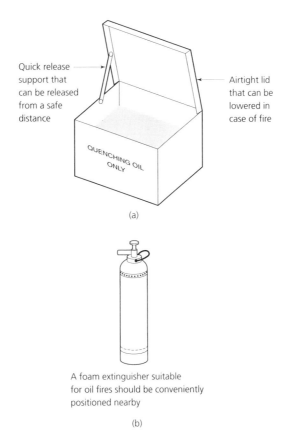

Quick release support that can be released from a safe distance

Airtight lid that can be lowered in case of fire

QUENCHING OIL ONLY

(a)

A foam extinguisher suitable for oil fires should be conveniently positioned nearby

(b)

All personnel working in heat-treatment shops must be alert to the possibility of fires. They must be conversant with and trained in the correct fire drill as laid down in the works notices, as shown in Fig. 7.7(a):

- Fire drills must be practised regularly to ensure that every one knows what to do.
- They must know where the nearest alarm is and how to operate it, as shown in Fig. 7.7(b).
- They must know where the nearest telephone is and how to summon the fire brigade, as shown in Fig. 7.7(c).
- They must know where the exits are and how to evacuate the premises. Fire exits must be left clear of obstructions, as shown in Fig. 7.7(d).
- They must know where to assemble so that a roll-call can be taken.
- They must know the correct fire-fighting procedures so that the fire can be contained until the professional brigade arrives. This should only be done, however, if it can be done safely.
- When an outbreak of fire is discovered, the correct fire drill must be actioned immediately.

Fig. 7.7 *Safety in the event of a fire: (a) works safety notices must be understood and obeyed; (b) know where the nearest alarm is and how to operate it; (c) know where the nearest phone is; (d) fire exits and notices must be kept clear of obstructions*

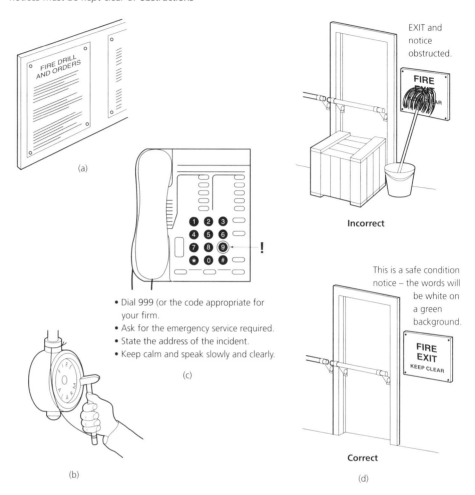

7.3 The heat treatment of plain carbon steels (reasons and processes)

The reasons for modifying the properties of plain carbon steels by heat treatment are as follows:

- To improve the properties of the material as a whole.
- To improve the functional properties of part of the materials – for example, hardening the surface of the material to prevent wear whilst leaving the core of the material relatively soft and tough.

- To remove undesirable properties acquired from previous working – for example, hardness induced by cold working.

The heat-treatment processes we are now going to consider in order to modify the properties of plain carbon steels are as follows:

- Through hardening.
- Quenching, distortion and cracking.
- Tempering.
- Annealing.
- Normalising.
- Case hardening.

7.3.1 *Through hardening*

Basically the process by which we 'through harden' (also referred to as quench hardening) consists of heating a suitable steel to a critical temperature and cooling it quickly (quenching it) in water or quenching oil. The hardness attained will depend upon:

- The carbon content of the steel (the higher the carbon content the harder the steel).
- The rate of cooling (the faster the cooling the harder the steel).

Figure 7.8 shows the stages in hardening a simple chisel made from a 0.8 per cent carbon steel.

Fig. 7.8 *Hardening a plain carbon steel chisel: (a) heat to cherry red; (b) quench in water*

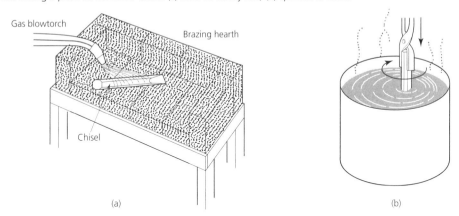

The temperature to which the steel is heated in this example should be between 800 and 830 °C – that is, the steel should glow 'cherry red' in colour. Any increase in temperature will make no difference to the hardness of the chisel; it will only result in 'grain growth' and weakening of the steel. Grain growth means that the hotter the steel becomes and the longer it is kept at excessively high temperatures the bigger the grains in the metal will grow by merging together. For strength and good machining properties, a fine-grain structure is required.

It has been stated previously that the steel has to be heated to within a critical temperature range if it is to be correctly hardened, but where do we find these critical temperatures? They are to be found in the steel section of the iron–carbon thermal equilibrium diagram. This is shown in Fig. 7.9.

Fig. 7.9 *Iron–carbon equilibrium diagram*

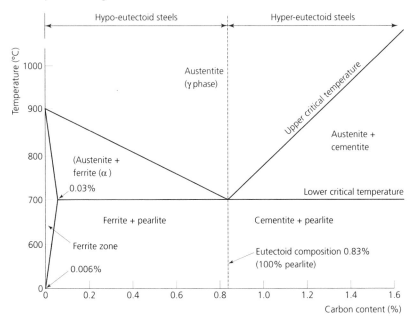

Don't worry about the name, all you need to know is the general shape of the diagram and the fact that the lines represent *critical points* (*temperatures*) for plain carbon steels with different carbon contents. These are the temperatures at which the structure of the steel changes. For your interest, I will briefly explain the meaning of the different terms used in the diagram.

- *Ferrite* is virtually pure iron (it contains only 0.006 per cent carbon at room temperature. It is very soft and ductile
- *Cementite* is iron and carbon chemically combined to form the compound iron carbide. Cementite is just another name for iron carbide, it is very hard and brittle.
- *Pearlite* grains consist of alternate layers (laminations) of ferrite and cementite. It is the toughest structure that can occur in a plain carbon steel.
- *Austenite* is a *solid solution* of carbon in iron. It cannot exist below the lower critical point (temperature) for plain carbon steels.
- *The eutectoid point* occurs at a carbon content of about 0.83 per cent. Only pearlite is present, so this is the toughest of the plain carbon steels. Pearlite changes instantly to austenite at the lower critical point (temperature) on heating, and back again on cooling.
- *Hypo-eutectoid* steels are those steels with a carbon content below 0.83 per cent. The name literally means 'below the eutectoid'.

- *Hyper-eutectoid steels* are those steels with a carbon content above 0.83 per cent. The name literally means 'above the eutectoid'.

So, how does this help us? Figure 7.10 shows the hardening temperatures for plain carbon steels. You will see that the hardening temperatures lie in a narrow band closely related to the critical points (temperatures). If the temperature range for any given steel is not achieved, it will not harden. If it is exceeded, grain growth will occur and the steel will be weakened.

Fig. 7.10 *Hardening temperatures for plain carbon steel*

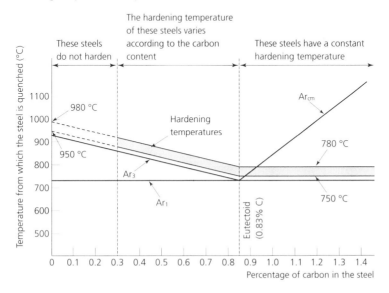

It has already been stated that the more quickly a component is cooled the harder it becomes for any given carbon content. However, some care is required because the faster the workpiece is cooled, the more likely it is to crack and distort. Therefore, the workpiece should never be cooled more quickly than is required to give the desired degree of hardness. The most common substances used for quenching (*quenching media*) are:

- *Brine* (salt water solution) This is the most rapid quenching bath – it will give the greatest hardness and is the most likely to cause cracking.
- *Water* This is less severe and is the most widely used quenching bath for plain carbon steels.
- *Oil* This is the least severe of the liquid quenching media. Only plain carbon steels of the highest carbon content will harden in oil, and then only in relatively small sections. Oil quenching is mostly used with alloy die steels and tool steels. Compared with water and brine it is also the most dangerous of the quenching media that is used. If it becomes heated above its flash point, it can ignite and burn. The safety precautions to be taken when using oil for quenching were explained in Section 7.2.
- *Air blast* This is the least severe of any of the quenching media used. It can only be applied to heavily alloyed steels of small section such as high-speed steel tool bits.

Table 7.2 summarises the effect of carbon content and rate of cooling for a range of plain carbon steels.

Table 7.2 _Effect of carbon content and rate of cooling on hardness_

Type of steel	Carbon content (%)	Effect of heating and quenching (rapid cooling)
Low carbon	Below 0.25	Negligible
Medium carbon	0.3–0.5	Becomes tougher
	0.5–0.9	Becomes hard
High carbon	0.9–1.3	Becomes very hard

Carbon content (%)	Quenching bath	Required treatment
0.30–0.50	Oil	Toughening
0.50–0.90	Oil	Toughening
0.50–0.90	Water	Hardening
0.90–1.30	Water	Hardening

Notes:
1. Below 0.5% carbon content, steels are not hardened as cutting tools, so water hardening has not been included.
2. Above 0.9% carbon content, any attempt to harden the steel in water could lead to cracking.

Overheating plain carbon steels

It is a common mistake to overheat a steel in the hope that it will become harder. As previously stated, the hardness only depends upon the carbon content and the rate of cooling. Once the correct hardening temperature has been reached, any further increase in temperature only slows up the time taken to cool the workpiece and this slower cooling only results in reduced hardness.

Overheating, particularly prolonged overheating, causes _grain growth_ which results in a weak and defective component. This condition can be corrected by normalising and rehardening. Normalising will be dealt with later in this chapter. If the overheating is excessive, then 'burning' occurs. This is oxidation of the grain boundaries of the metal resulting in great weakness which, unlike moderate overheating, cannot be corrected. The component can only be melted down as scrap. On the other hand, failure to reach the correct hardening temperature results in the workpiece remaining soft no matter how quickly it is quenched.

7.3.2 _Quenching and cracking_

Quenching and distortion

Quenching is not just a question of dunking the hot metal into the quenching bath like a biscuit into a cup of tea. Some thought must be given to the way the work is lowered into the bath to avoid distortion and get the most effective quenching. If you look back to

Fig. 7.8, you will see that the shank of the chisel is held in the tongs and that the chisel is dipped vertically into the quenching bath.

- This results in the cutting end of the steel entering the bath first and attaining maximum hardness whilst the quenching medium is at its minimum temperature.
- The shank is masked to some extent by the tongs and this results in reduced hardness. This does not matter in this example as we want the shank to be tough rather than hard so that it does not shatter when struck with a hammer.
- The chisel should be stirred around in the bath so that it is constantly coming into contact with fresh and cold water. It also prevents steam pockets being generated round the chisel which would slow the cooling rate and prevent maximum hardness being achieved.
- Dipping a long slender workpiece like a chisel vertically into the bath prevents distortion. This is shown in Fig. 7.11. This figure also shows what happens if the component is quenched flat and also how the shape of the component itself can cause uneven cooling and distortion.

Fig. 7.11 *Causes of distortion: (a) distortion caused by an unbalanced shape being hardened; (b) how to quench long, slender components to avoid distortion*

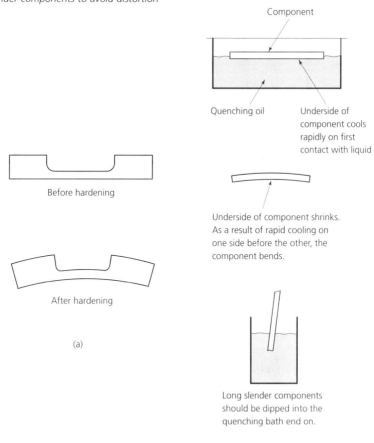

Component

Quenching oil

Underside of component cools rapidly on first contact with liquid

Before hardening

Underside of component shrinks. As a result of rapid cooling on one side before the other, the component bends.

After hardening

(a)

Long slender components should be dipped into the quenching bath end on.

(b)

Cracking

Figure 7.12 shows some typical causes of cracking occurring during, and as a result of, heat treatment. Careful design and the correct selection of materials can result in fewer problems in the hardening shop.

- Avoid sharp corners and sudden changes of section.
- Do not position holes, slots and other features near the edge of the workpiece.
- Do not include screw threads in a hardened component. Apart from the chance of cracking occurring, once hardened you cannot run a die down the thread to ease it if it has become distorted during the hardening process.
- For complex shapes, which are always liable to cracking and distortion during hardening, always use an alloy die steel that has been formulated so that minimum distortion and shrinkage (movement) will occur during heat treatment. Such steels are oil or air hardening and this also reduces the chance of cracking.

Fig. 7.12 *Causes of cracking: (a) incorrect engineering that promotes cracking; (b) correct engineering to reduce cracking*

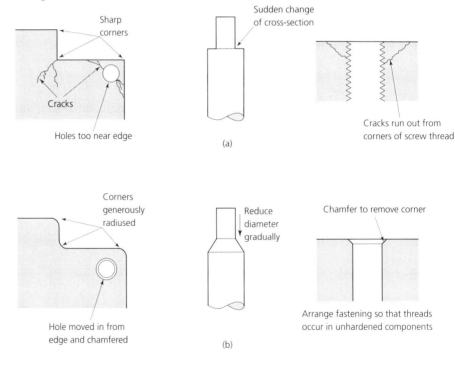

7.3.3 *Tempering*

When you heat and quench a plain carbon steel as described previously, you not only harden the steel, you also make it *very brittle*. In this condition it is unsuitable for immediate use. For instance, a chisel would shatter if you hit it with a hammer. After hardening we have to carry out another process known as *tempering*. This greatly reduces the brittleness and increases the toughness. However, the tempering process also reduce the hardness to some extent.

Tempering consists of reheating the hardened steel workpiece to a suitable temperature and quenching it in oil or water. The tempering temperature to which the workpiece is reheated depends only upon the use to which the workpiece is to be put. Table 7.3 lists some suitable temperatures for tempering components made from plain carbon steels.

Table 7.3 *Tempering temperatures*

Component	Temper colour	Temperature (°C)
Edge tools	Pale straw	220
Turning tools	Medium straw	230
Twist drills	Dark straw	240
Taps	Brown	250
Press tools	Brownish-purple	260
Cold chisels	Purple	280
Springs	Blue	300
Toughening (crankshafts)	—	450–600

In a workshop the tempering temperature is usually judged by the 'temper colour' of the oxide film that forms on the surface of the workpiece. After hardening, the surface of the workpiece is polished so that the colour of the oxide film can be clearly seen. Figure 7.13 shows a chisel being tempered. The chisel is not uniformly heated. As shown, the shank is heated in the flame and the temper colours are allowed to 'run down' the chisel until the cutting edge reaches the required colour. When the cutting edge is the required temper colour, the chisel is immediately 'dipped' vertically into the quenching bath again (see Fig. 7.8(b)). This gives the cutting edge the correct temper but leaves the shank softer and tougher so that it can withstand being struck with a hammer.

Fig. 7.13 *Tempering a cold chisel*

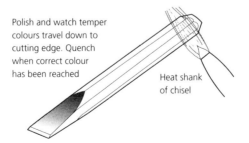

Polish and watch temper colours travel down to cutting edge. Quench when correct colour has been reached

Heat shank of chisel

Where uniform tempering is required – for example, a press-tool die – then the workpiece can be heated on a sand tray, as shown in Fig. 7.14. When the correct temper colour has been reached, the die is immediately dipped, edge-ways on, into the quenching bath. It must never be dipped in the flat position or it will distort as described previously. Complex and large components and batches of components should be tempered in a furnace with atmosphere and temperature control to ensure consistent results. Such furnaces will be described later in this chapter.

Fig. 7.14 *Tempering a die uniformly*

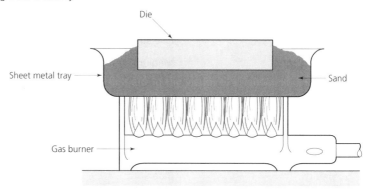

7.3.4 *Annealing*

Annealing processes are used to soften steels that are already hard. This hardness may be imparted in two ways.

- *Quench hardening*   This has previously been described in the section on 'Through hardening'.
- *Work hardening*   This occurs when the metal has been cold worked (see Section 6.2). It becomes hard and brittle at the point where cold working occurs as this causes the grain structure to deform. For example, if a strip of metal is held in a vice, bending the metal back and forth causes it to work harden at the point of bending. It will eventually become sufficiently hard and brittle to break off at that point.

Full annealing

Full annealing is carried out by heating the workpiece to the same temperatures that were used for hardening, and Fig. 7.15 shows the temperatures for full annealing and the temperatures for subcritical annealing. To *anneal* (soften) the workpiece, you allow the hot metal to cool down as slowly as possible. Small components can be buried in crushed limestone or in ashes. Larger components and batches of smaller components will have been heated in furnaces. When the correct temperature has been reached and the component is 'soaked' at this temperature so that the temperature becomes uniform throughout its mass. The furnace is then shut down, the flue dampers are closed and the furnace is sealed so that it cools down as slowly as possible with the work inside it.

Fig. 7.15 *Annealing temperatures for plain carbon steels*

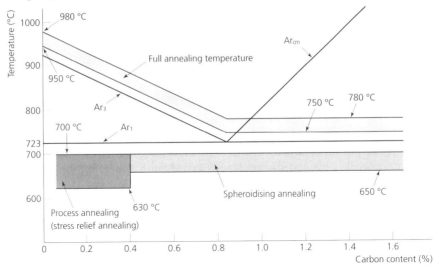

Although such slow cooling results in some grain growth and weakening of the metal, it will impart maximum ductility. This results in the metal being in the correct condition for cold forming. However, because of its extreme softness and grain growth the metal will tend to tear and leave a poor surface finish if it is machined. Components to be machined should be *normalised* as described below.

Stress-relief annealing

This process is reserved for steels with a carbon content below 0.4 per cent. Such steels will not satisfactorily quench harden but, as they are relatively ductile, they will be frequently cold worked and become work hardened. Since the grain structure will have become severely distorted by the cold working, the crystals will begin to reform and the metal will begin to soften at 500 °C. In practice, the metal is rarely so severely stressed as to trigger *recrystalisation* (see Section 6.2) at such a low temperature. Therefore, stress-relief annealing is usually carried out between 630 and 700 °C to speed up the process and prevent excessive grain growth. Stress-relief annealing has a variety of names.

- *Stress-relief annealing* since the metal is relieved of the stresses caused by previous forming processes.
- *Subcritical annealing* since heating is limited to temperatures below (sub-) the lower critical point (see Fig. 7.15).
- *Process annealing* since the work hardening of the metal results from cold-working (forming) processes.
- *Inter-stage annealing* since the process is often carried out between the stages of a process when extensive cold working is required – for example, when deep drawing sheet metal in a press.

The degree of stress-relief annealing and the rate of cooling will depend not only upon the previous processing the steel received before annealing, but also upon the processing it is to receive *after* annealing. If further cold working is to take place, then maximum softness and ductility is required. This is achieved by prolonged heating and very slow cooling to encourage grain growth. However, if grain refinement, strength and toughness are more important, then heating and cooling should be more rapid.

Spheroidising annealing

This is subcritical annealing applied to steels whose carbon content is greater than 0.4 per cent. Heating is to subcritical temperatures, usually between 650 and 700 °C as shown in Fig. 7.15. The iron carbide (cementite) tends to 'ball-up' into tiny *spheroids*. This would normally result in impaired physical properties and poor machining qualities. However, if this process is carried out after work hardening or, preferably, quench hardening, the steel will cold work and also machine to a good finish. Moreover, steels that have been subjected to spheroidising annealing will reharden more uniformly and with less chance of cracking and distortion. As with any other annealing process, slow cooling is required after heating.

7.3.5 *Normalising*

Plain carbon steels are normalised by heating them to the temperatures shown in Fig. 7.16. Note how, again, the heat-treatment temperatures are related to the iron-carbon thermal equilibrium diagram. This time the processing temperatures follow the lines connecting the upper critical point for the steel to ensure that, no matter what its carbon content, the steel will become austenitic and any coarse-grain structures in the steel will be destroyed. This is important because this time we want a finer grain structure in the steel. To achieve this, we have to heat the metal up more quickly and cool it more quickly. The workpiece is taken out of the furnace and allowed to cool in the free air of the heat-treatment shop. The air should be able to circulate freely round the workpiece. However, the workpiece must be sited so that it is free from cold draughts. Warning notices that the steel is dangerously hot must be placed around it.

Fig. 7.16 *Normalising temperatures for plain carbon steels*

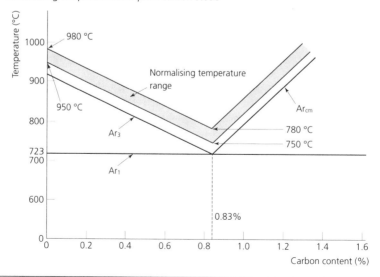

Castings and forgings are produced by high-temperature processes. As they cool down and shrink they often develop high internal stresses. Machining tends to partially release these stresses, and over a period of time the machined components 'move' or distort and become inaccurate. At one time, large castings and forgings were rough machined and left out of doors to 'weather' over a period of a year or more. The continual changes in temperature released all the stresses and the components became 'stabilised' ready for finish machining. No further movement then took place.

Keeping such a large amount of valuable stock tied up over a long period of time is no longer economically viable and weathering has given way to *normalising*. The normalising process is now frequently used for stress relieving between the rough machining of castings and forgings and the finish machining of such workpieces. This is done to stabilise such workpieces and to avoid 'movement' or distortion subsequent to machining. When normalising is used for the stabilisation of castings and forgings, it is sometimes referred to as 'artificial weathering'.

7.3.6 *Case hardening*

Often, components need to be hard and wear resistant on the surface, yet have a tough and strong core to resist shock loads. These two properties do not exist in a single steel. For toughness you require a steel whose carbon content does not exceed 0.3–0.4 per cent. For hardness you require a steel with a carbon content of about 1.0 per cent.

The usual solution to this problem is *case hardening*. This is a process by which carbon is added to the surface layers of low-carbon steels or low-alloy steels to a carefully regulated depth. This addition of carbon is called *carburising*. After carburising the component is put through successive heat-treatment processes to harden the case and refine the core. This process has two distinct steps, as shown in Fig. 7.17. First, the workpiece is heated to between 900 and 950 °C in contact with the carburising compound until the additional carbon has been absorbed to the required depth. Second, the workpiece is removed from the carburising compound, reheated to between 780 and 820 °C and dipped off (quenched) in cold water.

Fig. 7.17 *Case-hardening: (a) carburising; (b) after carburising; (c) after quenching component from a temperature above 780 °C*

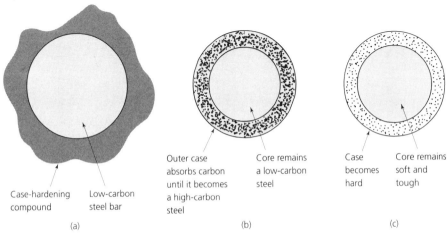

Carburising

This depends upon the fact that very low-carbon (0.1 per cent) steels heated to the austenitic condition will absorb carbon (check back with Fig. 7.9). Various carbonaceous materials are used in the carburising process:

- *Solid media*, such as bone charcoal or charred leather, together with an energiser such as sodium and/or barium carbonate. The energiser makes up to 40 per cent of the total composition.
- *Molten salts*, such as sodium cyanide, together with sodium carbonate and/or barium carbonate and sodium or barium chloride. Since cyanide is a deadly poison such salts must be handled with great care and the cyanide must only make up to between 20 and 50 per cent of the total. Stringent safety precautions must be taken in its use. The components to be carburised are immersed in the molten salts.
- *Gaseous media* based upon natural gas (methane) are increasingly used. Methane is a hydrocarbon gas containing organic carbon compounds that are readily absorbed into the steel. The methane gas is frequently enriched by the vapours that are given off when mineral oils are 'cracked' by heating them in contact with the metal platinum which acts as a catalyst.

It is a common fallacy that carburising hardens the steel. *It does not*, it only adds carbon to the surface of the steel and leaves the steel in a fully annealed (soft) condition. It is the subsequent heat treatment that hardens the steel.

Superficial hardening

This produces a shallow case on simple components in the workshop, as shown in Fig. 7.18.

Fig. 7.18 *Superficial hardening: (a) heat to bright red in brazing hearth; (b) plunge red-hot component into carburising powder (repeat to give required depth of case); (c) reheat to cherry red and quench in water*

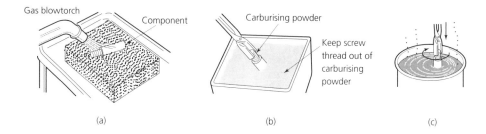

- The component is raised to red heat using a gas torch and brazing hearth.
- The red-hot component is then plunged into a case-hardening powder such as 'Kasenit'. This consists of carbon-rich compounds plus an energiser as previously described.
- The component absorbs the powder into its surface. This 'carburises' the surface of the metal and increases its carbon content.

- The heating and dipping can be repeated several times to increase the depth of carbon infusion.
- Finally, the component is again heated to red heat and plunged immediately into cold water. If the case-hardening powder has done its job, there should be a loud 'crack' and any surplus powder breaks away from the surface of the metal.
- The surface of the metal should now have a mottled appearance and should be hard. The component is now case hardened.
- Because this technique only results in a fairly shallow case, it is referred to as 'superficial hardening'. The case is not deep enough for finishing by grinding processes, although polishing is permissible.
- Bright-drawn steels do not absorb carbon readily unless the drawn surface is removed by machining. Wherever possible use a case-hardening quality steel which is formulated and processed to suit this treatment.

Deep case hardening

Where a deep case is required so that components can be finished by grinding (e.g. surface grinding or cylindrical grinding) the following procedure is required:

- The component is 'pack carburised' by burying it in the carburising compound in a steel box, sealed with an airtight lid.
- The box is heated in a furnace for several hours, depending upon the depth of case required. The carburising temperature is about 950 °C, as shown in Fig. 7.19.
- The box is removed from the furnace and allowed to cool. The component is removed from the box and cleaned in order to remove any residual powder from its surface.
- The component will be soft and have a coarse-grain structure because of the length of time it has been heated at a high temperature, and its subsequent slow cooling.

Fig. 7.19 *Case-hardening temperatures*

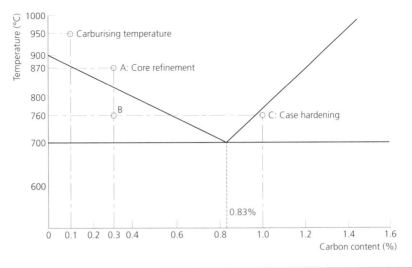

- The core of the component will have a carbon content of less than 0.3 per cent. The low-carbon steel core of the component is toughened by refining its grain. To do this, the component is heated to 870 °C as shown at A in Fig. 7.19 and then quenched in water.
- Since this temperature is well below the carburising temperature which caused the grain growth, and because of the rapid cooling, the core will now have a fine grain.
- This rapid cooling will also have the effect of hardening the case. Unfortunately the case will have a coarse grain since it was heated to above its correct hardening temperature, as shown at B in Fig. 7.19.
- To refine the grain of the case and reharden it, the component is heated to 760 °C, as shown at C in Fig. 7.19, and quenched in water. This is the correct hardening procedure for a 1.0 per cent carbon steel, which is what the surface of the component has become. This temperature is too low to affect the fine grain of the core.
- Finally the component can be tempered if required (see 'Tempering' page 250).

In place of a solid carburising compound, gas carburising or salt-bath carburising may be used. These media have been described previously.

Localised case hardening

It is often not desirable to case harden a component all over. For example, it is undesirable to case harden screw threads. Not only would they be extremely brittle, but any distortion occurring during carburising and hardening could only be corrected by expensive thread-grinding operations. Various means are available for avoiding local infusion of carbon during the carburising process.

- Heavily copper plating those areas to be left soft. The layer of copper prevents intimate contact between the component and the carburising medium. Copper plating cannot be used for salt-bath carburising as the molten salts dissolve the copper.
- Encasing the areas to be left soft in fire-clay. This technique is mostly used when pack carburising.
- Leaving surplus metal where a soft area is required. This is machined off between carburising and hardening (dipping off). An example is shown in Fig. 7.20. Although more expensive because of the extra handling involved, it is the most certain and effective way of leaving local soft features.

Fig. 7.20 *Localised case-hardening*

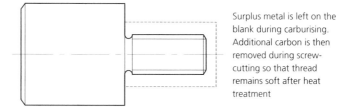

Surplus metal is left on the blank during carburising. Additional carbon is then removed during screw-cutting so that thread remains soft after heat treatment

Flame hardening

Localised surface hardening can also be achieved in medium- and high-carbon steels and some cast irons by rapid local heating and quenching. Figure 7.21(a) shows the principle of flame hardening. A carriage moves over the workpiece so that the surface is rapidly heated by an oxyacetylene or oxypropane flame. The same carriage carries the water-quenching spray. Thus the surface of the component is heated and quenched before the core material of the component can rise to its hardening temperature. This process is often used for hardening the slideways of machine tools – for example, the slideways on lathe beds. This process is also called the 'Shorter process' after the name of its inventor.

Fig. 7.21 *Surface hardening: (a) flame hardening (Shorter process); (b) induction hardening*

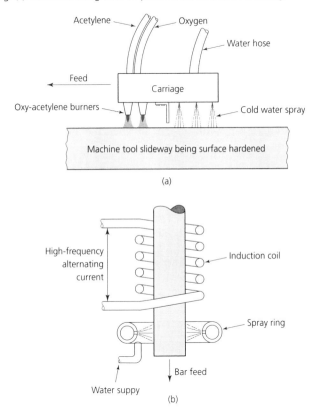

(a)

(b)

Induction hardening

Figure 7.21(b) shows how the same surface-hardening effect can be produced by high-frequency electromagnetic induction. The induction coil surrounding the component is connected to a high-frequency alternating current generator. This induces high-frequency eddy currents in the surface of the component causing it to become very hot, very rapidly. (This is similar to the action of a microwave oven.) When the hardening temperature has been reached, the current is switched off and a water spray quenches the component. The induction coil is made from perforated copper tube which also carries the quenching oil or water.

This technique is often used for hardening gear teeth. The induction coil can be tailored to suit the profile of the component. The use of high-frequency alternating current allows a smaller induction coil to be used and allows the effect to be more localised. Alternating currents always tend to travel along the surface of a conductor. The higher the frequency, the nearer to the surface will be the eddy currents that cause the heating. Therefore, the depth of the case can be controlled by the frequency of the generator.

Nitriding

This process puts an ultra-hard, wear-resistant coating onto components. The components are made from a special alloy steel containing either 1.0 per cent aluminium or traces of molybdenum, chromium and vanadium. Nitrogen gas is absorbed into the surface of the metal to form very hard metal nitrides. The process consists of heating the components in ammonia gas at about 500 °C for upwards of 40 hours. At this temperature the ammonia gas breaks down and the atomic nitrogen is readily absorbed into the surface of the steel. No quenching is required. The component is finished before nitriding since no subsequent grinding is possible as the nitrided layer is only a few micrometres thick. This is of no importance since the process does not affect the surface finish and the process temperature is too low to cause distortion. The advantages of this process can be summarised as:

- Cracking and distortion are eliminated since the processing temperature is relatively low and there is no subsequent quenching.
- Surface hardnesses as high as 1150 H_D are attainable with 'Nitralloy' steels. This is much harder than the results that can be attained by conventional hardening.
- The corrosion resistance of the steel is improved.
- The treated components retain their hardness up to an operating temperature of 500 °C, whereas conventionally hardened plain carbon and low-alloy steels start to soften at 220 °C.

7.4 The heat treatment of non-ferrous metals and alloys

None of the non-ferrous metals, and only a very few non-ferrous alloys, can be quench hardened like plain carbon steels. The majority of non-ferrous metals can only be hardened by cold working. Alternatively, they can be manufactured from cold-rolled (spring temper) sheet or strip, or they can be manufactured from cold-drawn wire. Work-hardened non-ferrous metals can be annealed by a recrystallisation process similar to the subcritical annealing for plain carbon steels. The main difference is that non-ferrous metals do not have to be cooled slowly. They can be quenched after heating, and this has the advantage that the rapid cooling causes the metal to shrink suddenly, removing the oxide film. This is even more effective if quenching takes place in a very weak solution of sulphuric acid. (*Safety*: If an acid bath is used, protective clothing and eye protection such as goggles or, better still, a visor **must** be worn.) Suitable annealing temperatures are:

- Aluminium   500–550 °C   (pure metal)
- Copper   650–750 °C   (pure metal)
- Cold-working brass   600–650 °C   (simple alloy of copper and zinc)

Heat-treatable aluminium alloys ('duralumin' is such and alloy) require somewhat different treatment. They can be softened by *solution treatment* and hardened by *natural ageing* or they can be hardened artificially by *precipitation treatment*. The alloy 'duralumin' contains traces of copper, magnesium, manganese and zinc; aluminium makes up the remainder of the alloy.

7.4.1 *Solution treatment*

To soften the alloy, it is raised to a temperature of about 500 °C (depending upon the alloy). At this temperature the alloying elements can form a solid solution in the aluminium. The alloy is quenched from this temperature to preserve the solution at room temperature. Gradually, the solid solution will break down with age and the alloy will become harder and more brittle. Therefore, solution treatment must be carried out immediately before the alloy is to be processed. The breakdown of the solution can be delayed by refrigeration at between $-6°$ and -10 °C. Conversely, it can be speeded up by raising the temperature.

7.4.2 *Precipitation treatment*

The natural hardening mentioned above is called *age hardening*. This is the result of hard particles of aluminium–copper intermetallic compounds precipitating out of the solid solution. This hardens and strengthens the alloy but makes it less ductile and more brittle. Precipitation hardening can be accelerated by heating the alloy to about 150–170 °C for a few hours. This process is referred to as *artificial ageing* or *precipitation hardening*. The times and temperatures vary for each alloy and the alloy manufacturer's heat-treatment specifications must be carefully observed, especially for critical components such as those used in the aircraft industry.

7.5 Heat-treatment furnaces

The requirements of heat-treatment furnaces are as follows:

- *Uniform heating of the work*   This is necessary in order to prevent distortion of the work due to unequal expansion, and also to ensure uniform treatment.
- *Accurate temperature control*   We have previously discussed the critical nature of heat-treatment temperatures. Therefore, not only must the furnace be capable of operating over a wide range of temperatures, but it must be easily adjustable to the required process temperature.
- *Temperature stability*   Not only is it essential that the temperature is accurately adjustable but, once set, the furnace must remain at the required temperature. This is achieved by ensuring that the mass of the heated furnace lining (refractory) is very much greater than the mass of the work (charge). It can also be achieved by automatic temperature control, or by both.

- *Atmosphere control* If the work is heated in the presence of air, the oxygen in the air attacks the surface of the metal to form metal oxides (scale). This not only disfigures the surface of the metal, it can also change the composition of the metal at its surface. For example, in the case of steels, the oxygen can also combine with the carbon at the surface of the metal. Reducing the carbon content results in the metal surface becoming less hard and/or tough. Furthermore, if the products of combustion of gas or oil fuels contain sulphur compounds, then these products of combustion can, on coming into contact with the work, increase the sulphur content of the workpiece material causing embrittlement and loss of strength. To provide atmosphere control, the air in the furnace is replaced with some form of inert gas which will not react with the workpiece material. Alternatively, the work may be totally immersed in hot, molten salts.
- *Economical use of fuels* This is essential if heat-treatment costs are to be kept to a minimum. If the furnaces can be run continuously on a shift work basis considerable economies can be made. The fuel required to keep firing up furnaces from cold is much greater than that required for continuous running. Furthermore, the continual expansion and contraction of the furnace lining resulting from intermittent use causes cracking and early failure of the furnace lining. Thus it is more economical for small workshops to contract their heat treatment out to specialist firms who have sufficient volume of work to keep their furnaces in continuous use.
- *Low maintenance costs* The furnace is lined with a heat-resistant material such as firebrick. Heat-resistant materials are called refractories. Since the furnace must be taken out of commission each time this lining is renewed, it should be designed to last as long as possible. In this connection, there is no economy in running a furnace at or beyond its recommended maximum operating temperature for any length of time. This only wastes fuel and causes an early breakdown of the lining.

Open hearth furnace

Figure 7.22 shows the simplest form of furnace. This is little different to heating a component on a brazing hearth, except that the furnace arch reflects heat back onto the component and provides rather more uniform heating. In this furnace a gas or oil burner plays directly onto the charge. Heat is also reflected onto the work from the furnace lining, as previously mentioned. The advantages and limitations of this type of furnace are as follows:

Advantages
- Low initial cost.
- Simplicity in use and maintenance.
- Fuel economy since it heats up quickly.

Limitations
- Uneven heating.
- Poor temperature control.
- Poor temperature stability.
- Complete lack of atmosphere control resulting in heavy scaling and flue gas contamination of the work.

Fig. 7.22 *Gas-heated open hearth furnace*

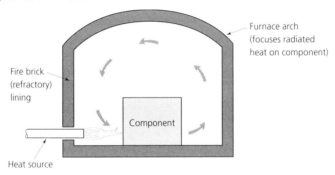

Semi-muffle furnace

Figure 7.23 shows a semi-muffle furnace. This is an improvement upon the open hearth furnace previously described. The flame from the burner does not play directly onto the charge, but passes under the hearth to provide 'bottom heat'. This results in more uniform heating The advantages and limitations of this type of furnace are as follows:

Advantages
- Comparatively low initial cost.
- Simplicity in use and maintenance.
- Fuel economy.
- Fairly rapid heating.
- Heating is more uniform than for the open hearth type of furnace.
- Limited atmosphere control can be achieved by varying the gas–air mixture through a system of dampers. The flue outlets are situated just inside the furnace door so that any atmospheric oxygen that may leak past the door is swept up the flue before it can add to the scaling of the work.

Fig. 7.23 *Gas-heated semi-muffle furnace*

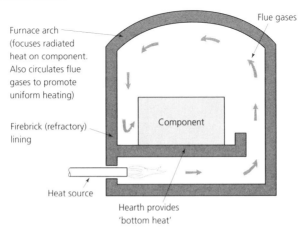

- Reasonable temperature control.
- Reasonable atmosphere stability due to the greater mass of the furnace lining compared with the open hearth type furnace.

Limitations
- Heating is still comparatively uneven compared with more sophisticated furnace types.
- Atmosphere control is somewhat limited. Although oxidation can be reduced by careful control of the gas–air mixture, some scaling will still take place and there will be still be flue gas contamination of the work.

Muffle furnace (gas heated)
Figure 7.24 shows a full muffle furnace. You can see from the figure that the work is heated in a separate compartment called a *muffle*. The work is completely isolated from the flame and the products of combustion. The advantages and limitations for this type of furnace are as follows:

Advantages
- Uniform heating.
- Reasonable temperature control.
- Good temperature stability due to the high mass of refractory material forming the muffle and the furnace lining compared with the mass of the work.
- Full atmosphere control is possible. Any sort of atmosphere can be maintained within the muffle since no combustion air is required in the muffle chamber.

Limitations
- Higher initial cost.
- Maintenance more complex and costly.
- Greater heat losses and slow initial heating results in lower fuel economy unless the furnace can be operated continuously.

Fig. 7.24 *Gas-heated muffle furnace*

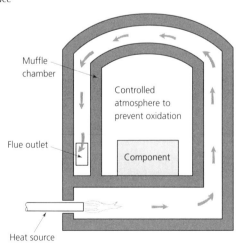

Muffle furnace (electric resistance)

Figure 7.25 shows a typical electric resistance muffle furnace. The electric heating elements are similar to those found in domestic electric ovens. They are independent of the atmosphere in which they operate. Therefore they can be placed within the muffle chamber itself. This results in a higher operating efficiency compared with the gas-heated muffle furnace and more than offsets the higher energy cost for electricity compared with gas. The advantages and limitations of this type of furnace are as follows:

Advantages
- Uniform heating of the work.
- Accurate temperature control.
- Ease of fitting automatic control instrumentation.
- High temperature stability.
- Full atmosphere control.
- Comparatively easy maintenance.

Limitations
- Higher energy source costs.
- Lower maximum operating temperatures, as above 950–1000 °C the life of the resistance elements is low.

Fig. 7.25 *Electrically heated muffle furnace; (a) the electric resistance furnace; (b) heating element*

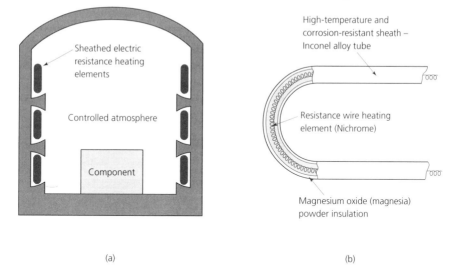

(a)   (b)

Muffle furnace (electric induction)

The principles of electromagnetic induction heating were described in 'Induction hardening' on page 259. These same principles can be applied to a muffle furnace similar to the one shown in Fig. 7.25, except that the resistance heater is replaced by an induction coil. Since the heating currents are induced in the work itself, the induction coil does not have to be inside the muffle chamber. It can be wrapped around the outside of the firebrick lining of the muffle chamber and is, therefore, unaffected by the heat of the work. This enables the work to be heated to very much higher temperatures without affecting the life of the induction coil. Such furnaces are suitable for the hardening of high-alloy tool steels such as high-speed steel. As with all muffle furnaces, full atmosphere control can be provided.

Salt-bath furnace (gas fired)

Figure 7.26 shows a typical gas-fired salt-bath furnace. Points to note are:

- *Tangential firing*   This ensures that the flame does not play directly on the pot.
- *Top heat*   In the interests of safety, the salts must be heated from the top downwards. If heated from the bottom, the expanding molten salts erupt through the crust like a miniature volcano. Since the molten salts could be red hot, this would be very dangerous.
- *Preheating*   All work, and baskets for containing the work, must be preheated to ensure that they are perfectly dry. Moisture entering the molten salts can cause an explosion that throws the molten salts out of the pot.

Fig. 7.26   *Gas-heated salt-bath furnace*

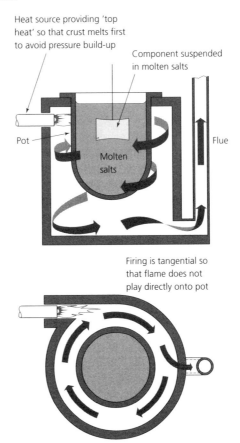

Heat source providing 'top heat' so that crust melts first to avoid pressure build-up

Component suspended in molten salts

Pot

Flue

Molten salts

Firing is tangential so that flame does not play directly onto pot

The salts used are dependent upon the process being carried out. All reputable suppliers will recommend the most suitable salts to suit a particular situation:

- *Nitrate-based salts* These are used for low-temperature applications such as the tempering of ferrous metals and the solution treatment of aluminium alloys. If overheated these salts can explode violently and special sections of the Health and Safety at Work, etc., Act cover their use. Automatic temperature control with fail-safe features are essential.
- *Chloride-based salts* These are used for the general heat treatment of ferrous metals – for example, quench hardening, annealing and normalising. Nitrate-based salts would be used for tempering as stated above.
- *Cyanide-based salts* These are used for case hardening low-carbon and alloy steel components. Since these salts are fatally poisonous, extreme care must be exercised in their use and disposal. Special sections of the Health and Safety at Work, etc., Act govern their use.

Note: It is usual to use an 'economiser' in the form of mica flakes floated on the surface of the molten salts to prevent loss through oxidation and fuming.

The advantages and limitations of salt-bath furnaces are as follows:

Advantages
- Absolute uniformity of heating as the work is enveloped in molten salt at the treatment temperature.
- Accurate temperature control.
- High-temperature stability if the mass of the work is small compared with the mass of the molten salts.
- No atmosphere control is required as the charge is enveloped in the molten salts.

Limitations
- Low fuel economy unless used on a continuous basis because of the time taken to heat up the furnace and melt the salts.
- Regular maintenance required.
- Salt baths are potentially dangerous owing to the risk of eruption of the salts if the work is damp and the risk of an explosion if the salts are overheated. Also some of the salts used are poisonous. Thus, a relatively highly trained workforce is required to operate salt-bath furnaces.

Salt-bath furnace (electrically heated)
You can see from Fig. 7.27 that in this type of furnace two electrodes are immersed in the salts. These pass a heavy current through the salts at a low voltage. The resistance of the salts to the passage of the current causes them to heat up rapidly and become molten. Although electricity is more expensive than gas, the fact that heating occurs within the salts themselves renders this type of furnace highly efficient and economically comparable with gas- and oil-fired furnaces. Electric heating is readily adaptable to automatic control where high-temperature stability is required, as in the heat treatment of alloy steels, or in the use of nitrate salts. The general advantages and limitations are the same as for gas-heated salt-bath furnaces.

Fig. 7.27 *Electrically heated salt-bath furnace*

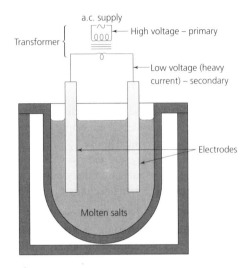

7.6 Temperature measurement

The importance of temperature measurement and control during heat-treatment processes has already been discussed in this chapter. The familiar mercury-in-glass thermometer is inadequate at the temperatures of most heat-treatment processes. However, the mercury-in-steel thermometer and the mercury vapour-pressure thermometer are suitable for the lower temperature processes used for tempering and the solution treatment of light alloys. For higher temperatures one or other of the temperature-measuring devices known as *pyrometers* are required.

7.6.1 *Mercury-in-steel thermometer*

The principle parts of this device are shown in Fig. 7.28. A stainless-steel bulb is connected to a pressure gauge by a fine-bore (capillary) tube of steel. The whole system is filled with liquid mercury so that no air or mercury vapour is present. As the bulb is heated, the mercury expands faster than the steel and increases the pressure acting on the pressure gauge. This is calibrated in degrees of temperature so that direct temperature measurements can be taken. Providing the mass of mercury in the connecting tube is small compared with the mass of mercury in the bulb, the connecting tube can be up to 30 metres in length. The limiting temperature is about 600 °C, at which the mercury commences to vaporise.

Fig. 7.28 *Mercury-in-steel thermometer*

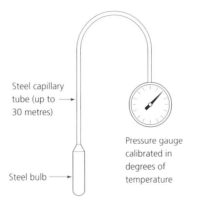

Steel capillary
tube (up to
30 metres)

Pressure gauge
calibrated in
degrees of
temperature

Steel bulb

7.6.2 *Vapour-pressure thermometer*

The basic construction of this type of thermometer is the same as that shown in Fig. 7.28, except that only the bulb contains mercury. The rest of the instrument contains mercury vapour. When the bulb is heated more mercury vaporises and it is the increase in mercury vapour pressure that actuates the pressure gauge. Again, the gauge is calibrated in degrees of temperature. Providing the connecting tube has a fine bore, it can be up to 30 metres in length. The limiting temperature for this type of thermometer is about 800 °C, but it is not as sensitive or as accurate as the mercury-in-steel thermometer at lower temperatures. The relatively low cost, simplicity and reliability of the *mercury-in-steel* and the *vapour-pressure* thermometers makes them very attractive for the lower temperature heat-treatment processes.

7.6.3 *Thermocouple pyrometer*

This is the most widely used temperature-measuring device for heat-treatment purposes. Figure 7.29(a) shows the principle of the thermocouple pyrometer. If the junction of two wires made from dissimilar metals (such as a copper wire and an iron wire) form part of a closed electric circuit and the junction is heated, a small electric current will flow. The presence of this current can be indicated by a sensitive galvanometer. Increasing the temperature difference between the hot and cold junctions increases the current in the circuit. If the galvanometer is calibrated in degrees of temperature, we have a temperature-measuring device called a *pyrometer*. Figure 7.29(b) shows how these principles can be applied to a practical thermocouple pyrometer. The component parts of this instrument are:

- The thermocouple probe (hot junction).
- The indicating instrument (milli-ammeter).
- The 'ballast' or 'swamp' resistor.
- The compensating leads.

Fig. 7.29 *The thermocouple pyrometer: (a) principle of operation; (b) pyrometer circuit; (c) thermocouple probe*

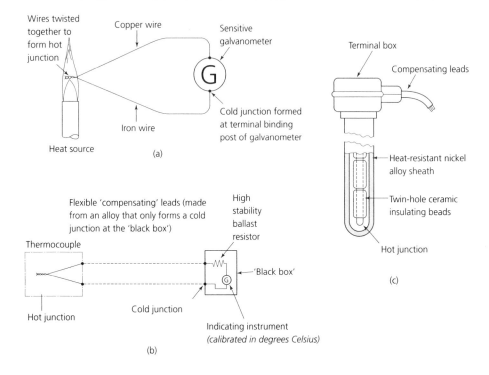

Thermocouple probe

This consists of a junction of two wires of dissimilar metals contained within a tube of refractory metal or of porcelain. Porcelain beads are used to insulate the two wires and locate them in the sheath, as shown in Fig. 7.29(c). Table 7.4 lists the more usual hot junction material combinations, together with their temperature ranges and sensitivities.

Table 7.4 *Thermocouple combinations*

Thermocouple	Sensitivity (millivolts/°C)	Temperature range (°C)
Copper–constantan	0.054	−220 to +300
Iron–constantan	0.054	−220 to +750
Chromel–alumel	0.041	−200 to +1200
Platinum–platinum/rhodium	0.0095	0 to +1450

Notes:
Constantan = 60% copper, 40% nickel
Chromel = 90% nickel, 10% chromium
Alumel = 95% nickel, 2% aluminium, 3% manganese
Platinum/rhodium = 90% platinum, 10% rhodium

Indicating instrument

This is a sensitive milli-ammeter calibrated in degrees Celsius (°C) so that a direct reading of temperature can be made. A common error is to set this instrument to read *zero* when the system is cold. In fact it should be set to read the *atmospheric temperature* at the point of installation. The terminals of this instrument form the cold junction and should be placed in a cool position where they are screened from the heat of the furnace.

The 'ballast' or 'swamp' resistor

This is contained within the case of the indicating instrument. Its purpose is to give stability to the system. The resistance of electrical conductors increase as their temperature increases and the conductors that make up a pyrometer circuit are no exception. The variation in resistance with temperature would seriously affect the calibration of the instrument if the ballast resistor was not present. This resistor is made from manganin wire. (Manganin is an alloy whose resistance is virtually unaffected by heat.) By making the resistance of the ballast resistor very large compared with the resistance of the rest of the circuit, it *swamps* the effects of any changes in resistance that may occur in the rest of the circuit and renders them unimportant.

Compensating leads

These are used to connect the thermocouple probe to the indicating instrument. They are made of a special alloy so that they form a cold junction with the terminals of the indicating instrument, but not with the terminals of the probe. To avoid changes in calibration, the compensating leads must not be changed in length, nor must alternative conductors be used. The thermocouple, compensating leads and the indicating instrument must always be kept together as a set.

7.6.4 *The radiation pyrometer*

This device is used to measure the temperature:

- Of large hot components that have been removed from the furnace.
- Where the furnace temperature is so high it would damage the thermocouple probe.
- Where the hot component is inaccessible – where the temperature of the component in the furnace needs to be measured rather that the temperature of the furnace atmosphere itself.

The principle of this type of pyrometer is shown in Fig. 7.30. Instead of the thermocouple probe being inserted into the furnace atmosphere, the radiant heat from the furnace or the component being heated in the furnace is focused onto the thermocouple by a parabolic mirror.

You must remember that as the temperature of a work reaches the furnace temperature, the rate at which the temperature of the work increases becomes less. It is difficult to assess just when, if ever, the component reaches furnace temperature. Certainly, the soaking time involved would give rise to excessive grain growth. Furnaces are frequently operated above the required temperature, and the work is withdrawn from the furnace when it has reached its correct temperature as measured by a radiation pyrometer.

Fig. 7.30 *The radiation pyrometer*

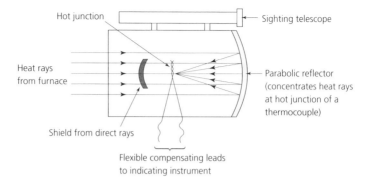

For the production heat treatment in, say, a steel works, where large coils of rolled steel strip are to be process annealed, the elapsed time alone can be used. We can assume that all the coils are of the same mass and the same material. Each coil represents the furnace charge and the furnace is run at a higher temperature than the temperature required to heat treat the charge. The charge is loaded into the furnace and the time it takes to reach the process temperature, as indicated on a radiation pyrometer, is noted. Provided that no change takes place in the temperature of the furnace or the mass of the charge, then each new charge will be correctly treated if it is heated for the time originally recorded.

7.6.5 *Temperature assessment*

The devices and techniques described above give precise temperature measurement. There are simpler ways of assessing the *approximate* temperature some of these will now be described.

Paints and crayons

These are applied to the surface of the component to be heat treated. The mark left on the surface by their application changes in colour and appearance when the desired temperature has been reached. The paints and crayons are available in a range of compositions to suit the temperature required. They have the advantage of indicating the temperature of the component at the point of application. It has been stated previously that the temperature of the charge does not necessarily reflect the temperature in the furnace. Paints and crayons can also be used to indicate the preheating temperature of components to be joined by welding. Another application is to mark the blades of gas turbines (jet engines) so that, when undergoing routine maintenance, it can be seen if the blades have been overheated and therefore weakened.

Ceramic cones

These are also known as 'Seger' cones and may be conical or pyramidal in shape. The latter have a triangular base. The 'cones' are made with various compositions so that they soften at different temperatures. It is usual to choose three cones, one slightly below the required temperature (cone A in Fig. 7.31), one at the required temperature (cone B in Fig. 7.31), and one slightly above the required temperature (cone C in Fig. 7.31).

- If the furnace is below the required temperature, none of the cones softens and droops, as shown in Fig. 7.31(a).
- If the furnace is too hot, all the cones will droop as shown in Fig. 7.31(b).
- If the furnace is at the correct temperature, cone A will droop a lot, cone B will just start to droop at the tip, and cone C will be unaffected. This situation is shown in Fig. 7.31(c).

Fig. 7.31 *Use of Seger cones: (a) temperature too low – no cones soften and droop; (b) temperature too high – all the cones soften and droop; (c) temperature correct – cone A softens and droops, cone B just starts to droop, cone C is unaffected*

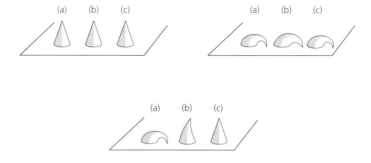

7.7 Atmosphere control

When natural gas is burnt in a furnace, excess air is usually present to ensure complete and efficient combustion. The resulting *products of combustion* (flue gases) contain oxygen, carbon dioxide, sulphur, nitrogen and water vapour. These all react to a greater or lesser degree with the surface of the workpiece whilst it is in the furnace. They will produce heavy scaling and, in the case of steel, surface decarburisation and softening. The situation is not so serious in the case of a muffle furnace as the fuel is burnt in a separate chamber and cannot come into contact with the work. However, the oxygen and water vapour in the air are still present in the muffle chamber and will cause some scaling and decarburisation.

To eliminate this effect, the air in the muffle chamber can be replaced by alternative atmospheres, depending upon the process being performed and the metal being treated. This is known as *atmosphere control*. These controlled atmospheres are based upon natural gas (methane) and LPG gases such as propane and butane. For special applications, ammonia gas and 'cracked' ammonia gas are used. Where complete freedom from contaminants is required, *vacuum* furnaces can be used. These have to be specially manufactured so that the furnace does not 'implode' as a result of the forces exerted by the external atmospheric pressure. The work is put into the furnace chamber which is then sealed. The air and any residual water vapour are pumped out to create a vacuum and only then does heating commence. Electric induction heating is generally used for this type of furnace.

7.8 Summary of safety procedures

Various safety requirements associated with heat treatment have been mentioned throughout this chapter. Because of their importance they will now be summarised:

- The two most common hazards when heat treating metals are burns, which can be serious and painful, and fire.
- Operators should be provided with appropriate protective clothing such as flame-retardent or flame-resistant overalls and caps, leather gloves, leather aprons, leather spats, industrial safety shoes and goggles or a transparent face visor.
- Always wear the clothing provided.
- Assume that everything is hot until you have proved it is cold. Warning notices should be put adjacent to large and hot workpieces.
- Oil quenching baths should have an airtight lid. If the oil catches fire, closing the lid stops the air feeding the flames and puts the fire out. Leave the lid on until the oil cools down.
- Oil quenching baths should be provided with a circulating and cooling system to keep the temperature of the oil below its flash point, at which it ignites. This is particularly important when production quenching large quantities of work.
- Never use lubricating oil for quenching. It is too flammable and gives off clouds of noxious fumes. Use only properly formulated quenching oils.
- Never light up or shut down a furnace until you have been properly instructed and given permission.
- Make sure the furnace has been properly ventilated (particularly if it has not been used for some time) so that there is no possibility of it containing a build up of fuel gas that could cause an explosion.
- Open all the dampers so that there is adequate secondary combustion air present.
- Reduce the primary air supply to the burners to prevent them backfiring (popping back) when lighting up. Once lit, the primary air can be increased to give the combustion pattern required.
- Never use a salt-bath furnace without first drying and preheating the work and work baskets before putting them into the salts.
- Never tamper with temperature-measuring and controlling devices.
- Never use copper wire to hang work in the salts. Some salts dissolve copper and the work will drop into the salts and may crack the crucible, allowing the red-hot molten salts to flood out.
- Learn where the fire extinguishers are kept and how to use them.
- Learn the safety drill in case of fire. Read the safety notices.
- Learn what to do if a workmate's clothing catches fire.
- Learn how to give the alarm if a fire breaks out.

7.1 *Safety*
   (a) Briefly describe the type of clothing and protective devices you should wear when carrying out heat-treatment processes.
   (b) Sketch **three** warning signs you would expect to find in a heat-treatment shop.

7.2 *Reasons for heat treatment*
   (a) State the main **two** purposes for the heat treatment of metallic materials.
   (b) Explain why a coppersmith would anneal a blank cut from a sheet of copper before beating it to shape, and why he/she would need to reanneal the metal from time to time as forming proceeds.

7.3 *Hardening plain carbon steels*
   (a) What two factors does the hardness of a plain carbon steel depend upon when through hardening?
   (b) Explain why steels have to be tempered after hardening and how the degree of temper is controlled when this is done over a brazing hearth in the workshop.
   (c) When through hardening, state the effect of:
      (i) overheating the steel
      (ii) underheating the steel
   (d) When through hardening, explain how the hot metal should be quenched and state the precautions that must be taken to avoid cracking and distortion.

7.4 *Local hardening*
   (a) With the aid of sketches, show how a simple component can be superficially case-hardened at the brazing hearth in a workshop.
   (b) List the operations for case hardening the component shown in Fig. 7.32 so that the threads are left soft.

Fig. 7.32 *Exercise 7.4(b). (Dimensions in millimetres)*

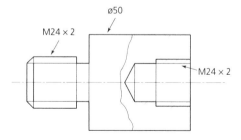

   (c) Name an example of:
      (i) a solid case-hardening compound
      (ii) a liquid case-hardening compound
      (iii) a gaseous case-hardening compound

7.5 *Annealing and normalising*

(a) Describe the essential differences between annealing and normalising plain carbon steels.

(b) Describe the essential differences between full-annealing and subcritical annealing as applied to plain carbon steels.

(d) Describe the essential differences between the annealing of plain carbon steels and the annealing of non-ferrous metals (other than the heat-treatable aluminium alloys).

(e) Described how 'duralumin' is softened. What is the name of the process used, and what is the name of natural process by which this aluminium alloy gradually becomes hard again?

7.6 *Heat-treatment equipment*

(a) List the main requirements of a heat-treatment furnace.

(b) With the aid of sketches, describe any heat-treatment furnace with which you are familiar. Draw particular attention to its main features. List the main advantages and limitations for the furnace type chosen.

(c) Describe the precautions that must be taken when starting up and shutting down furnaces.

(d) Describe the need for, and a method of, atmosphere control in heat-treatment furnaces.

(e) Describe one low-technology and one high-technology method of temperature measurement suitable for a furnace used for the occasional hardening of high-carbon steel components.

8 General health and safety (engineering)

When you have read this chapter, you should understand:

- The statutory requirements for general health and safety at work.
- Accident and first aid procedures.
- Fire precautions and procedures.
- Protective clothing and equipment.
- Correct manual lifting and carrying techniques.
- How to use powered lifting equipment.
- Safe working practices.

8.1 Health, safety and the law

8.1.1 *Health and Safety at Work, etc., Act*

It is essential to observe safe working practices not only to safeguard yourself, but also to safeguard the people with whom you work. The Health and Safety at Work, etc., Act provides a comprehensive and integrated system of law for dealing with the health, safety and welfare of work-people and the general public as affected by industrial, commercial and associated activities. The Act has six main provisions:

- To completely overhaul and modernise the existing law dealing with safety, health and welfare at work.
- To put general duties on employers, ranging from providing and maintaining a safe place to work, to consulting on safety matters with their employees.
- To create a Health and Safety Commission.
- To reorganise and unify the various Government Inspectorates concerned with industrial safety.
- To provide powers and penalties for the enforcement of safety laws.
- To establish new methods of accident prevention, and new ways of operating future safety regulations.

The Act places the responsibility for safe working equally upon:

- The employer.
- The employee (that means you).
- The manufacturers and suppliers of materials, goods, equipment and machinery.

8.1.2 *Health and Safety Commission*

The Act provides for a full-time, independent chairman and between six and nine part-time commissioners. The commissioners are made up of three trade union members appointed by the TUC, three management members appointed by the CBI, two local authority members, and one independent member. The commission has taken over the responsibility previously held by various government departments for the control of most occupational health and safety matters. The commission is also responsible for the organisation and functioning of the Health and Safety Executive.

Health and Safety Executive

This unified inspectorate combines together the formerly independent Government Inspectorates such as the Factory Inspectorate, the Mines and Quarries Inspectorate, and similar bodies. Since 1975 they have been merged together into one body known as the Health and Safety Executive Inspectorate. The inspectors of the HSE have wider powers under the Health and Safety at Work, etc., Act than under previous legislation and their duty is to implement the policies of the Commission.

Should an inspector find a contravention of one of the provisions of earlier Acts or Regulations still in force, or a contravention of the Health and Safety at Work, etc., Act, the inspector has three possible lines of action available.

Prohibition Notice
If there is a risk of serious personal injury, the inspector can issue a *Prohibition Notice*. This immediately stops the activity giving rise to the risk until the remedial action specified in the notice has been taken to the inspector's satisfaction. The prohibition notice can be served upon the person undertaking the dangerous activity, or it can be served upon the person in control of the activity at the time the notice is served.

Improvement Notice
If there is a legal contravention of any of the relevant statutory provisions, the inspector can issue an *Improvement Notice*. This notice requires the infringement to be remedied within a specified time. It can be served on any person on whom the responsibilities are placed. The latter person can be an employer, employee or a supplier of equipment or materials.

Prosecution
In addition to serving a Prohibition Notice or an Improvement Notice, the inspector can prosecute any person (including an employee – you) contravening a relevant statutory provision. Finally, the inspector can seize, render harmless or destroy any substance or article which the inspector considers to be the cause of imminent danger or personal injury.

Thus every employee must be a fit and trained person capable of carrying out his or her assigned task properly and safely. Trainees must work under the supervision of a suitably trained, experienced worker or instructor. By law, every employee must:

- Obey all the safety rules and regulations of his or her place of employment.
- Understand and use, as instructed, the safety practices incorporated in particular activities or tasks.
- Not proceed with his or her task if any safety requirement is not thoroughly understood, in which case guidance must be sought.
- Keep his or her working area tidy and maintain his or her tools in good condition.
- Draw the attention of his or her immediate supervisor or the safety officer to any potential hazard.
- Report all accidents or incidents (even if injury does not result from the incident) to the responsible person.
- Understand emergency procedures in the event of an accident or an alarm.
- Understand how to give the alarm in the event of an accident or an incident such as fire.
- Cooperate promptly with the senior person in charge in the event of an accident or an incident such as fire.

Therefore, safety health and welfare are very personal matters for a young worker, such as yourself, who is just entering the engineering industry. This chapter sets out to identify the main hazards and suggests how they may be avoided. Factory life, and particularly engineering, is potentially dangerous and you must take a positive approach towards safety, health and welfare.

Further legislation and regulations concerning safety

In addition to the Health and Safety at Work, etc., Act, the following are examples of legislation and regulations that also control the conditions under which you work and the way in which you work (behaviour):

- Factories Act 1961
- Safety Representatives and Safety Committees Regulations 1977
- Notification of Accidents and General Occurrences Regulations 1980
- The Management of Health and Safety at Work Regulations 1992
- The Protection of Eyes Regulations 1974
- Electricity at Work Regulations 1989
- Low Voltage Electrical Equipment (Safety) Regulations 1989. (This includes voltage ranges of 50–1000 volts (AC) and 75–1500 volts (DC).)
- Abrasive Wheels Regulations 1970
- Noise at Work Regulations 1989

You are not expected to have a detailed knowledge of all this legislation, but you are expected to know of its existence, the main topic areas that it covers, and how it affects your working conditions, your responsibilities, and the way in which you work.

There are many other laws and regulations that you will come across depending upon the branch of engineering industry in which you work.

8.2 Employer's responsibilities

All employers must, by law, maintain a safe place to work. It is management's responsibility to implement a regular system of scheduled safety inspections for each workplace, process and machine. For certain items of plant, the frequency of inspection and the records to be kept are laid down in the relevant Regulations or are recommended in applicable Codes of Practice (e.g. cranes, lifting tackle, compressed air receivers, steam boilers, etc.). It is essential to establish clearly what equipment is to be inspected and by whom. Also, it is essential to establish clearly who is to supervise these inspection activities and what records of inspection are to be kept. In some instances outside professional inspectors will have to be called in. There should be a procedure for checking that corrective action is taken promptly following an adverse inspection report. Again, records must be kept of the inspection and the corrective action taken.

To fulfil all the obligations imposed upon them by current safety legislation, employers must ensure that:

- They have provided a safe place to work: safe access and exit so that in the case of an emergency (such as fire) no one will be trapped. This is particularly important when the workplace is not at ground level. Pedestrian access and exits should be segregated from lorries delivering materials or collecting finished work. The premises must be kept in good repair. Worn floor coverings and stair treads are a major source of serious falls.
- The plant and equipment are safe and that they comply with the *Machinery Directive* – that is, they are correctly installed and properly maintained. The plant and any associated cutters and tools must also be properly guarded.
- Working practices and systems are safe and that, where necessary, protective clothing is provided.
- A safe, healthy and comfortable working environment is provided. The temperature and humidity should be maintained at the correct levels for the work being undertaken.
- There is an adequate supply of fresh air and that fumes and dust are either eliminated all together or are reduced to an acceptable and safe level.
- There is adequate and suitable natural and artificial lighting, particularly over stairways.
- There is adequate and convenient provision for washing and sanitation.
- There is adequate first-aid facilities under the supervision of a qualified person. This can range from a first-aid box under the supervision of a person trained in basic first-aid procedures for a small firm, to a full-scale ambulance room staffed by professionally qualified medical personnel in a large firm.
- Provision is made for the safe handling, storing, siting and transportation of raw materials, work in progress and finished goods awaiting delivery.
- Provision is made for the safe handling, storing, siting, transportation and use of dangerous substances such as compressed gases (e.g. oxygen and acetylene), and toxic and flammable solvents.
- There is a correct and legal system for the reporting of accidents and the logging of such accidents in the *accident register*.
- There is a company policy for adequate instruction, training and supervision of employees. This must not only be concerned with safety procedures but also with good working practices. Such instruction and training to be updated at regular intervals.

- There is a safety policy in force. This safety policy must be subject to regular review. One of the more important innovations of the Health and Safety at Work, etc., Act is contained in section 2(4) which provides for the appointment of safety representatives from amongst the employees, who will represent them in consultation with the employers, and have other prescribed functions.
- Where an employer receives a written request from at least two safety representatives to form a *safety committee* the employer shall, after consulting with the applicants and representatives of other recognised unions (if applicable) whose members work in the workplace concerned, establish a safety committee within the period of three months after the request. The employer must post a notice of the composition of the committee and the workplaces covered. The notice must be positioned where it may be easily read by the employees concerned.
- Membership of the safety committee should be settled by consultation. The number of management representatives should not exceed the number of safety representatives. Where a company doctor, industrial hygienist or safety officer/adviser is employed they should be ex-officio members of the committee.
- Management representation should be aimed at ensuring the necessary knowledge and expertise to provide accurate information on company policy, production needs and technical matters in relation to premises, processes, plant, machinery and equipment.

8.3 Employee's responsibilities

All employees (including you) are as equally responsible for safety as their employers. Under the Health and Safety at Work, etc., Act, employees are expected to take reasonable care for their own health and safety together with the health and safety of other people with whom they work, and members of the public who are affected by the work being performed.

Further, the misuse of, or interference with, equipment provided by an employer for health and safety purposes is a criminal offence. It is up to all workers to develop a sense of *safety awareness* by following the example set by their instructors.

Regrettably not all older workers observe the safety regulations as closely as they should. Take care in whom you choose for your 'role model'. The basic requirements for safe working are to:

- Learn the safe way of doing each task. This is usually the correct way.
- Use the safe way of carrying out the task in practice.
- Ask for instruction if you do not understand a task or have not received previous instruction.
- Be constantly on your guard against careless actions by yourself or by others.
- Practice good housekeeping at all times.
- Cooperate promptly in the event of an accident or a fire.
- Report all accidents to your instructor or supervisor.
- Draw your instructor's or your supervisor's attention to any potential hazard you have noticed.

8.4 Electrical hazards

The most common causes of electrical shock are shown in Fig. 8.1. The installation and maintenance of electrical equipment must be carried out only by a fully trained and registered electrician. The installation and equipment must conform to international standards and regulations as laid down in safety legislation and the codes of practice and regulations published by the Institution of Electrical Engineers (IEE).

Fig. 8.1 *Causes of electric shock*

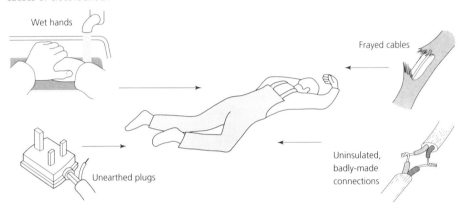

An electric shock from a 240 volt single-phase supply (lighting and office equipment) or a 415 volt three-phase supply (most factory machines) can easily kill you. Even if the shock is not sufficiently severe to cause death, it can still cause serious injury. The sudden convulsion caused by the shock can throw you from a ladder or against moving machinery. To reduce the risk of shock, all electrical equipment should be earthed or double insulated. The reason for earthing is shown in Fig. 8.2.

Further, portable power tools should be fed from a low-voltage transformer at 110 volts, as shown in Fig. 8.3. The machine must be suitable for operating at such a voltage. The transformer itself should be protected by a circuit breaker containing a residual current detector.

The fuses and circuit breakers designed to protect the supply circuitry to the transformer react too slowly to protect the user from electric shock. For this reason the supply to a portable power tool should also be protected by a residual current detector (RCD). Such a device compares the magnitudes of the current flowing in the live and neutral conductors supplying the tool. Any leakage to earth through the body of the user or by any other route will upset the balance between these two currents. This results in the supply being immediately disconnected. The sensitivity of residual current detectors is such that a difference of only a few milliamperes is sufficient to cut off the supply and the time delay is only a few microseconds. Such a small current applied for such a short time is not dangerous.

Fig. 8.2 *The need for earthing: (a) earth-fault-loop path; (b) the need for earthing metal-clad equipment*

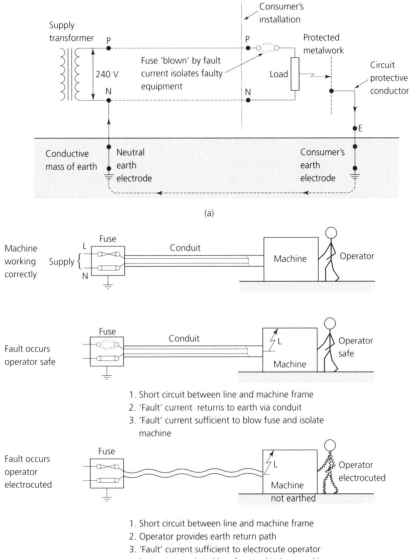

(a)

1. Short circuit between line and machine frame
2. 'Fault' current returns to earth via conduit
3. 'Fault' current sufficient to blow fuse and isolate machine

1. Short circuit between line and machine frame
2. Operator provides earth return path
3. 'Fault' current sufficient to electrocute operator but not enough to blow fuse and isolate machine

(b)

Fig. 8.3 *Low-voltage supply for portable power tools. (Note: Single-wound 'auto-transformers' do not provide isolation of the supply and must **not** be used)*

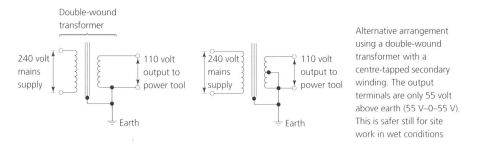

Double-wound transformer

240 volt mains supply — 110 volt output to power tool

Earth

240 volt mains supply — 110 volt output to power tool

Earth

Alternative arrangement using a double-wound transformer with a centre-tapped secondary winding. The output terminals are only 55 volt above earth (55 V–0–55 V). This is safer still for site work in wet conditions

In the event of rendering first aid to the victim of electrical shock, great care must be taken when pulling the victim clear of the fault which caused the shock. The victim can act as a conductor and thus, in turn, electrocute the rescuer. If the supply cannot be quickly and completely disconnected, always pull the victim clear by his or her clothing which, if dry, will act as an insulator. If in doubt, hold the victim with a plastic bag or cloth known to be dry. Never touch the victim's bare flesh until the victim is clear of the electrical fault. Artificial respiration must be started immediately the victim has been pulled clear of the fault or the live conductor. Figure 8.4 – reproduced by courtesy of the *Electrical Times* – gives details of how to treat a victim of severe electrical shock.

8.5 Fire fighting

Fire fighting is a highly skilled operation and most medium and large firms have properly trained teams who can contain the fire locally until the professional brigade arrives. The best way you can help is to learn the correct fire drill; both how to give the alarm and how to leave the building. It only requires one person to panic and run in the wrong direction to cause a disaster.

In an emergency never lose your head and panic.

Smoke is the main cause of panic. It spreads quickly through a building, reducing visibility and increasing the risk of falls down stairways. It causes choking and even death by asphyxiation. Smoke is less dense near the floor: as a last resort, *crawl*. To reduce the spread of smoke and fire, keep fire doors closed at all times but never locked. The plastic materials used in the finishes and furnishings of modern buildings give off highly toxic fumes. Therefore, it is best to leave the building as quickly as possible and leave the fire fighting to the professionals who have breathing apparatus. *Saving human life is more important than saving property*.

If you do have to fight a fire there are some basic rules to remember. A fire is the rapid oxidation (burning) of flammable materials at relatively high temperatures. Figure 8.5

Fig. 8.4 *Treatment for electric shock*

Order of action

1. **Switch off current**
   Do this immediately. If not possible do not waste time searching for the switch.

2. **Secure release from contact**
   Safeguard yourself when removing casualty from contact. Stand on non-conducting material (rubber mat, DRY wood, DRY linoleum). Use rubber gloves, DRY clothing, a length of DRY rope or a length of DRY wood to pull or push the casualty away from the contact.

3. **Start artificial respiration**
   If the casualty is not breathing artificial respiration is of extreme urgency. A few seconds delay can mean the difference between success or failure. Continue until the casualty is breathing satisfactorily or until a doctor tells you to stop.

4. **Send for doctor and ambulance**
   Tell someone to send for a doctor and ambulance immediately and say what has happened. Do not allow the casualty to exert himself by walking until he has been seen by a doctor. If burns are present, ask someone to cover them with a dry sterile dressing.

If you have difficulty in blowing your breath into the casualty's lungs, press his head further back and pull chin further up. If you still have difficulty, check that his lips are slightly open and that the mouth is not blocked, for example, by dentures. If you still have difficulty, try the alternative method, mouth-to-mouth, or mouth-to-nose, as the case may be.

Mouth-to-mouth

Lay casualty on back, if immediately possible, on a bench or table with a folded coat under shoulders to let head fall back. Kneel or stand by casualty's head. Press his head fully back with one hand and pull chin up with the other.

Breathe in deeply. Bend down, lips apart and cover casualty's mouth with your well open mouth. Pinch his nostrils with one hand. Breathe out steadily into casualty's lungs. Watch his chest rise.

Turn your own head away. Breathe in again.

Repeat 10 to 12 times per minute.

If the patient does not respond

 NORMAL PUPILS   DILATED PUPILS

Check carotid pulse, pupils of eyes and colour of skin.
   Pulse present, pupils normal – continue inflations until recovery of normal breathing (Steps 1, 2 and 3.)

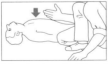

Pulse absent, pupils dilated, skin grey – strike smartly to the left part of breast bone with edge of hand.
   Response of continued pulse, pupils contract – continue inflations until recovery of normal breathing.

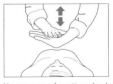

No response of continued pulse, pupils unaltered, skin grey – commence external heart compression.

When normal breathing commences, keep warm, place casualty in the recovery position.

External heart compression

1. Place yourself at the side of the casualty.

2. Feel for the lower half of the breastbone.

3. Place the heel of your hand on this part of the bone, keeping the palm and fingers off the chest.

4. Cover this hand with the heel of the other hand.

5. With arms straight, rock forwards pressing down on the lower half of the breastbone (in an unconscious adult it can be pressed towards the spine for about one and a half inches (4 cm)).

6. The action should be repeated about once a second

Continue as above until a continued pulse is felt and pupils contract.
Continue inflations until recovery of normal breathing

shows that removing the air (oxygen), or the flammable materials (fuel), or lowering the temperature will result in the fire ceasing to burn. It will go out. It can also be seen from Fig. 8.5 that different fires require to be dealt with in different ways.

Fig. 8.5 *How to remove each of the three items necessary to start a fire. (Note: Once the fire has started it produces sufficient heat to maintain its own combustion reactions and sufficient surplus heat to spread the fire)*

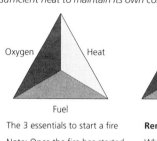

Oxygen   Heat

Fuel

The 3 essentials to start a fire	**Remove heat**	**Remove oxygen**	**Remove fuel**
Note: Once the fire has started it produces sufficient heat to maintain its own combustion reactions and sufficient surplus heat to spread the fire	When solids are on fire remove heat by applying water	Liquids, such as petrol etc. on fire can be extinguished by removing oxygen with a foam or dry powder extinguisher	Electrical or gas fires can usually be extinguished by turning off the supply of energy

8.5.1 *Fire extinguishers*

The normally available fire extinguishers and the types of fire they can be used for are as follows.

Water

Used in large quantities, water reduces the temperature and puts out the fire. The steam generated also helps to smother the flames as it displaces the air and therefore the oxygen essential to the burning process. However, for various technical reasons, water should only be used on burning solids such as wood, paper and some plastics. A typical hose point and a typical pressurised water extinguisher is shown in Fig. 8.6.

Fig. 8.6 *Hose point (a) and pressurised water extinguisher (b)*

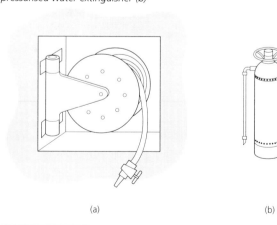

(a)   (b)

Foam extinguishers

These are used for fighting oil and chemical fires. The foam smothers the flames and prevents the oxygen in the air from reaching the burning materials at the seat of the fire. Water alone cannot be used because oil floats on water and this spreads the area of the fire. A typical foam extinguisher is shown in Fig. 8.7(a).

Note: Since both water and foam are electrically conductive, do not use them on fires associated with electrical equipment or the person wielding the hose or the extinguisher will be electrocuted.

Fig. 8.7 *Fire extinguishers: (a) foam; (b) CO_2; (c) vaporising liquid; (d) dry powder*

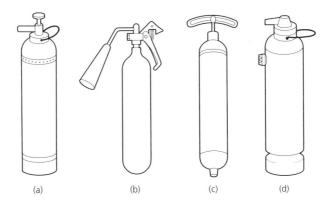

(a) (b) (c) (d)

Carbon dioxide (CO_2) extinguishers

These are used on burning gases and vapours. They can also be used for oil and chemical fires in confined places. The carbon dioxide gas replaces the air and smothers the fire. It can only be used in confined places, where it cannot be displaced by draughts.

Note: If the fire cannot breathe, neither can you, so care must be taken to evacuate all living creatures from the vicinity before operating the extinguisher. Back away from the bubble of CO_2 gas as you operate the extinguisher, do not advance towards it. Figure 8.7(b) shows a typical CO_2 extinguisher.

Vaporising liquid extinguishers

These include CTC, CBM and BCF extinguishers. The heat from the fire causes rapid vaporisation of the liquid sprayed from the extinguisher, and this vapour displaces the air and smothers the fire. Since a small amount of liquid produces a very large amount of vapour, this is a very efficient way of producing the blanketing vapour. Any vapour which will smother the fire will also smother all living creatures, which must be evacuated before using such extinguishers. As with CO_2 extinguishers, always back away from the bubble of vapour, never advance into it. Vaporising liquid extinguishers are suitable for oil, gas, vapour and chemical fires. Like CO_2 extinguishers, vaporising liquid extinguishers are safe to use on fires associated with electrical equipment. A typical example of a vaporising liquid extinguisher is shown in Fig. 8.7(c).

Dry powder extinguishers

These are suitable for small fires involving flammable liquids and small quantities of solids such as paper. They are also useful for fires in electrical equipment, offices and kitchens since the powder is not only non-toxic, but it can easily be removed by vacuum cleaning and there is no residual mess. The active ingredient is powdered sodium bicarbonate (baking powder) which gives off carbon dioxide when heated. A typical example of a dry powder extinguisher is shown in Fig. 8.7(d).

8.5.2 *General rules governing the use of portable extinguishers*

- Since fire spreads quickly, a speedy attack is essential if the fire is to be contained.
- Sound the alarm immediately the fire is discovered.
- Send for assistance before attempting to fight the fire.
- Remember:
    - (a) Extinguishers are only provided to fight small fires.
    - (b) Take up a position between the fire and the exit, so that your escape cannot be cut off.
    - (c) **Do not** continue to fight the fire if:
        - (i) it is dangerous to do so
        - (ii) there is any possibility of your escape route being cut off by fire, smoke, or collapse of the building
        - (iii) the fire spreads despite your efforts
        - (iv) toxic fumes are being generated by the burning of plastic furnishings and finishes
        - (v) there are gas cylinders or explosive substances in the vicinity of the fire.

If you have to withdraw, close windows and doors behind you wherever possible, but not if such actions endanger your escape. Finally, ensure that all extinguishers are recharged immediately after use.

8.6 Fire precautions and prevention

8.6.1 *Fire precautions*

It is the responsibility of employers and their senior management (duty of care) to ensure the safety of their employees in the event of fire. The following precautions should be taken:

- Ensure ease of exit from the premises at all times – emergency exits must not be locked or obstructed.
- Easy access for fire appliances from the local brigade.

- Regular inspection of the plant, premises and processes by the local authority fire brigade's fire prevention officer. No new plant or processes involving flammable substances should be used without prior notification and inspection by the fire prevention officer.
- The above note also applies to the company's insurance inspector.
- Regular and frequent fire drills must be carried out and a log kept of such drills, including the time taken to evacuate the premises. A roll call of all persons present should be taken immediately the evacuation is complete. A meeting of the safety committee should be called as soon as possible after a fire drill to discuss any problems, improve procedures and to learn lessons from the exercise.

8.6.2 *Fire prevention*

Prevention is always better than cure, and fire prevention is always better than fire fighting. Tidiness is of paramount importance in reducing the possibility of outbreaks of fire. Fires have small beginnings and as it is usually amongst accumulated rubbish that many fires originate you should make a practice of constantly removing rubbish, shavings, off-cuts, cans, bottles, waste paper, oily rags, and other unwanted materials to a safe place at regular intervals. Discarded foam plastic packing is not only highly flammable, but gives off highly dangerous toxic fumes when burnt.

Highly flammable materials should be stored in specially designed and equipped compounds away from the main working areas. Only minimum quantities of such materials should be allowed into the workshop at any time, and then only into *non-smoking zones*. The advice of the local authority fire brigade's fire prevention officer should also be sought.

It is good practice to provide metal containers with airtight hinged lids with proper markings as to the type of rubbish they should contain since some types of rubbish will ignite spontaneously when mixed. The lids of bins should be kept closed so that, if a fire starts in a bin, it it will quickly use up the air and go out of its own accord without doing any damage.

Liquid petroleum gases (LPG), such as propane and butane, are being used increasingly for process heating and for space heating in workshops and on site. Full and empty cylinders should be stored separately in isolated positions away from the working area and shielded from the sun's rays. The storage area should be well ventilated with ample space between the cylinders. The cylinders should also be protected from frost. Where bulk storage cylinders and spheres are used, not only must the above precautions be observed, but the containers must be securely fenced and defended against damage by passing vehicles.

Pipe run, joints and fittings associated with LPG installations must be regularly inspected by a suitably qualified and registered person. Flexible tubing (hoses) used in connection with gas cylinders must be regularly inspected for cuts and abrasions and replaced as necessary. The storage of oxygen and acetylene cylinders as used for welding require special treatment. The safety regulations associated with oxyacetylene welding and electric arc welding are dealt with in detail in *Fundamentals of Fabrication and Welding Engineering*.

8.7 Accidents

Accidents do not happen, they are caused. There is not a single accident which could not have been prevented by care and forethought on somebody's part. Accidents can and must be prevented. They cost millions of lost man-hours of production every year, but this is of little importance compared with the immeasurable cost in human suffering.

In every eight-hour shift nearly one hundred workers are the victims of industrial accidents. Many of these workers will be blinded, maimed for life, or confined to a hospital bed for months. At least two of them will die. Figure 8.8 shows the main causes of accidents.

Fig. 8.8 *Average national causes of industrial accidents (by per cent of all accidents)*

25.5 Handling and lifting goods and materials	**7.5** Struck by falling objects
19.4 Machinery	**7.5** Transport
15.9 Persons falling from heights or same level	**7.1** Use of hand tools
8.1 Stepping on or striking against objects	**9.0** Other causes, including electric shock

8.7.1 *Accident procedure*

You must learn and obey the accident procedures for your company.

- Report all accidents, no matter how small and trivial they seem, to your supervisor, instructor or tutor. Record your report and details of the incident on an accident form.

- Receive first-aid treatment from a *qualified* first aider, or your company's medical centre, depending upon the size of your company and its policy.

It is important that you follow the procedures laid down by your company since the accident register has to be produced on request by any HSE inspector visiting your company. Failure to log all accidents is an offence under the Health and Safety at Work, etc., Act and can lead to prosecution in the courts. Also, if at some future date you had to seek compensation as a result of the accident, your report is important evidence.

8.7.2 *Signs*

Warning signs and labels

You must be aware of the warning signs and their meanings. You must also obey such signs. To disregard them is an offence under the Health and Safety at Work, etc., Act. Warning signs are triangular in shape. All the sides have the same length of 1 metre. The background colour is *yellow* and there is a *black border* 67 mm wide. Figure 8.9 shows the 10 warning signs and gives their meanings.

Fig. 8.9 *Warning signs*

Caution, risk of fire	Caution, risk of explosion	Caution, toxic hazard	Caution, corrosive substance	Caution, risk of ionizing radiation

Caution, overhead load	Caution, industrial trucks	Caution, risk of electric shock	General warning, caution, risk of danger	Caution, laser beam

In addition to warning signs there are also *warning labels*. These should be affixed to all receptacles containing hazardous substances, and this applies both to raw materials used in manufacture and also to waste products. They are also used to indicate temporary hazards, such as machines that are under repair. Some typical hazard labels are shown in Fig. 8.10.

Prohibition signs

You can recognise these signs as they have a red circular band and a red cross-bar on a white background. Red should cover at least 35 per cent of the area of the sign and the

Fig. 8.10   *Warning labels (source: Warwick Sign and Display Ltd)*

symbol must not obscure the cross-bar. Figure 8.11 shows five typical prohibition signs. These signs indicate activities that are *prohibited* at all times. They *must be obeyed*, you have no option in the matter. To disregard them is an offence in law, as you would be putting yourself and others at considerable risk.

Fig. 8.11   *Prohibition signs*

| No smoking | Smoking and naked flames prohibited | Pedestrians prohibited | Do not extinguished with water | Not drinking water (that is, do not drink) |

Mandatory signs

You can recognise these signs as they have a blue background, which should cover at least 50 per cent of the area of the sign. The symbol must be white. Figure 8.12 shows five typical mandatory signs. These signs indicate things that *you must do* and precautions that *you must take*. These signs *must be obeyed*, you have no option in the matter. To disregard them is an offence in law as, again, you would be putting yourself at considerable risk.

Fig. 8.12   *Mandatory signs*

| Eye protection must be worn | Head protection must be worn | Hearing protection must be worn | Foot protection must be worn | Hand protection must be worn |

Safe condition signs

In addition to the signs discussed so far that tell you what to look out for, what you must do and what you must not do, there are also signs that tell what is safe. These have a white symbol on a green background. The green background should cover 50 per cent of the sign. The example shown in Fig. 8.13(a) indicates a first-aid post or an ambulance room. The example shown in Fig. 8.13(b) indicates a safe direction in which to travel – for instance, the safe direction towards an emergency exit. If Fig. 8.13(a) and (b) were used together they would indicate the correct and safe route to the first-aid post.

Fig. 8.13   *Safe condition signs*

First aid

(a)

Indication of direction

(b)

- Background colour shall be green.
- The symbol or text shall be white. The shape of the sign shall be oblong or square as necessary to accommodate the symbol or text.
- Green shall cover at least 50% of the area of the safety sign.

8.8 First aid

Accidents can happen anywhere and at any time. They can happen in the home and in the streets as well as in the workshops of industry. The results of such accidents can range from minor cuts and bruises to broken bones and life-threatening injuries, and it is a very good idea to know what to do in an emergency.

- You must be aware of the accident procedure.
- You must know where to find your nearest First Aid post.
- You must know the quickest and easiest route to the First Aid post.
- You must know who is the *qualified* first aider on duty (if he/she is a part-time first aider, you must know where he/she can be found).

All injuries are best dealt with by a *qualified* first aider. Unfortunately, in this day and age, more and more people are being encouraged to seek compensation through the courts of law. Complications resulting from amateurish but well-intentioned and well-meaning attempts on your part could result in you being sued for swingeing damages. Therefore, it is advisable to get some basic first-aid training in your own interests as well as those of your workmates. At least you will know your limitations and you will know what to leave alone for the professionally trained experts to deal with.

Best of all you should seriously consider becoming a fully qualified first aider by joining one of the following organisations in your spare time.

- St. John Ambulance Brigade.
- St. Andrew's Ambulance Association.
- British Red Cross Society.

These organisations jointly publish a comprehensive First Aid Manual to which the author is indebted for the following information. Although you can learn a lot from this manual, and you should keep a copy handy, it is no substitute for receiving proper training and certification under the supervision of one or other of the organisations listed earlier.

8.8.1 *Action in the event of an emergency*

If you are first on the scene of a serious incident, but you are not a trained first aider:

- Remain calm.
- Get help quickly by sending for the appropriate skilled personnel.
- Act and speak in a calm and confident manner to give the casualty confidence.
- Do not attempt to move the casualty.
- Do not administer fluids.
- Hand over to the experts as quickly as possible.

Giving First Aid can be dangerous. Always be watchful for your personal safety. It is possible to pick up serious virus infections such as hepatitis B or HIV if suitable precautions are not taken. Although there is a very small theoretical risk, there are no records of either of these diseases being passed on by mouth-to-mouth recusitation.

However, more serious is the possibility of blood-to-blood cross-contamination. Always wear disposable plastic or latex examination gloves when dealing with open wounds and blood. If you become concerned that you might have caught something whilst giving first aid, consult your doctor immediately. Regular first aiders should consider a course of immunisation followed by boosters at suitable intervals to protect against Hepatitis B.

8.8.2 *Minor wounds*

Prompt first aid can help nature heal small wounds and deal with germs. However, you must seek medical advice if:

- There is a foreign body embedded in the wound.
- There is a special risk of infection (such as a dog bite or the wound has been caused by a dirty object).
- A non-recent wound shows signs of becoming infected.

Figure 8.14 (a) and (b) shows the stages in dressing a wound involving minor bleeding. Minor bleeding is readily controlled by pressure and elevation. A small adhesive dressing is normally all that is necessary. Medical aid need only be sought if the bleeding does not stop or if the wound is at special risk of infection.

Fig. 8.14 *Minor wounds and bruises*

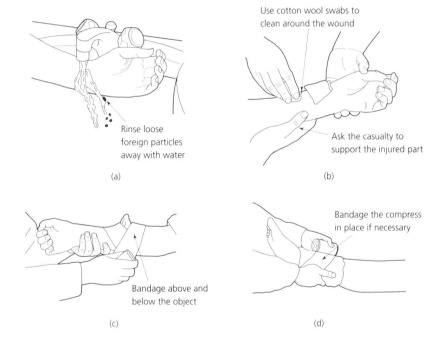

Rinse loose foreign particles away with water

(a)

Use cotton wool swabs to clean around the wound

Ask the casualty to support the injured part

(b)

Bandage above and below the object

(c)

Bandage the compress in place if necessary

(d)

1. Wash your hands thoroughly in soap and warm water.
2. If the wound is dirty, clean it by rinsing lightly under running water as shown in Fig. 8.14(a).
3. Pat gently dry with a sterile swab or clean tissue.
4. Temporarily cover the wound with sterile gauze. Clean the skin around the wound with soap and water (or a degreasing cleanser). Swab away from the wound and use a new swab for each stroke, as shown in Fig. 8.14(b).
5. Pat dry, then cover the wound with an adhesive dressing (plaster).

Note: If there is special risk of infection advise the casualty to consult his or her doctor. Avoid touching wounds with your bare fingers: wear disposable gloves. Don't talk, cough, sneeze or breathe over the wound or the dressings.

Sometimes there can be *foreign bodies* in minor wounds. Small pieces of glass or grit lying on a wound can be picked off with tweezers or rinsed off with cold water before treatment. However, you **must not** try to remove objects that are embedded in the wound; you may cause further tissue damage and bleeding.

1. Control any bleeding by applying firm pressure on either side of the object, and raising the wounded part.
2. Drape a piece of gauze lightly over the wound to minimise the risk of germs entering it, then build up padding around the object until you can bandage without pressing down upon it. If you cannot build up the padding high enough, bandage around the embedded foreign body as shown in Fig. 8.14(c).
3. Take or send the casualty to hospital.

Bruises

These are caused by internal bleeding that seeps through the tissues to produce the discoloration under the skin. Bruising may develop very slowly and appear hours, even days, after injury. Bruising that develops rapidly and seems to be the main problem will benefit from first aid. *Caution*: bruises may indicate deeper injury.

Treatment is aimed at reducing the blood flow to the injury, and minimising swelling by means of cooling and compression. If you suspect a more serious underlying injury, such as a sprain or fracture, seek medical advice.

1. Raise and support the injured part in a comfortable position.
2. Apply a cold compress to the bruise, as shown in Fig.8.14(d).

Minor burns and scalds

These are treated to stop the burning, to relieve pain and swelling and to minimise the risk of infection. If you are in any doubt as to the severity of the injury seek the advice of a doctor. The stages in treating minor burns and scalds are shown in Fig. 8.15.

Fig. 8.15 *Minor burns and scalds*

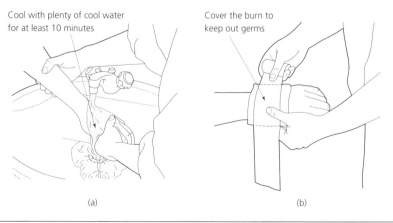

Cool with plenty of cool water for at least 10 minutes

Cover the burn to keep out germs

(a)

(b)

1. Flood the injured part with cold water for 10 minutes to stop the burning and relieve the pain. This is shown in Fig. 8.15(a). If water is unavailable, any cold harmless liquid such as milk or canned drinks will do.
2. Gently remove any jewellery, watches, or constricting clothing from the injured area before it begins to swell. Do not remove anything sticking to the burn.
3. Cover the area with a sterile dressing, or any clean, non-fluffy material, as shown in Fig. 8.15(b). A polythene bag or kitchen film makes a good temporary covering.

- **DO NOT** break blisters or interfere with the injured area; you are likely to introduce an infection.
- **DO NOT** use adhesive dressings or strapping.
- **DO NOT** apply lotions, ointments, creams or fats to the injury.

Note: Chemical burns to the skin, and particularly the eyes, require immediate and specialist treatment beyond the scope of this book. Expert attention must be obtained *immediately*.

Foreign bodies in the eye

Foreign bodies in the eye can lead to blurred vision with pain or discomfort. They can also lead to redness and watering of the eye, and the eyelid screwed up in spasm. A speck of dust or grit, or a loose eyelash floating on the white of the eye, can generally be removed easily. However, a foreign body that adheres to the eye, penetrates the eyeball, or rests on the coloured part of the eye should **not** be removed by a first aider. **Do not** touch anything sticking to, or embedded in, the eyeball or the coloured part of the eye. Cover the affected eye with an eye pad, then take or send the casualty to hospital.

First-aid treatment is aimed at preventing injury to the eye. The procedures are illustrated in Fig. 8.16.

Fig. 8.16 *Foreign bodies in the eye*

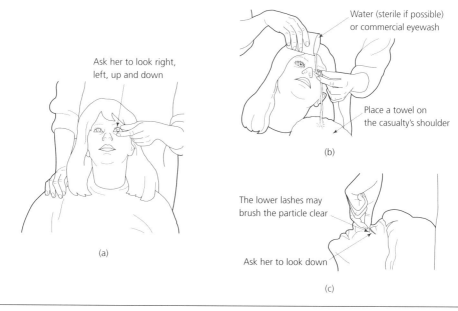

Ask her to look right, left, up and down

(a)

Water (sterile if possible) or commercial eyewash

Place a towel on the casualty's shoulder

(b)

The lower lashes may brush the particle clear

Ask her to look down

(c)

Advise the casualty not to rub his or her eye. Sit the casualty down facing the light. Gently separate the eyelids with your finger and thumb. Examine every part of the eye as shown in Fig. 8.16(a).

If you can see the foreign body, wash it out using a tumbler or an eye irrigator and clean water (sterile, if possible), as shown in Fig. 8.16(b). If this is unsuccessful then, providing the foreign body is not stuck in place, lift it off with a clean swab, or the damp corner of a tissue or a clean handkerchief.

If the object is under the eyelid, grasp the lashes and pull the upper lid over the lower lid as shown in Fig. 8.16(c). Blinking the eye under water may also make the foreign body float clear.

8.9 Personal protection

8.9.1 *Appearance*

Clothing

For general workshop purposes a boiler suit is the most practical and safest form of clothing. However, to be completely effective certain precautions must be taken, as shown in Fig. 8.17.

Fig. 8.17 *Correct and incorrect dress*

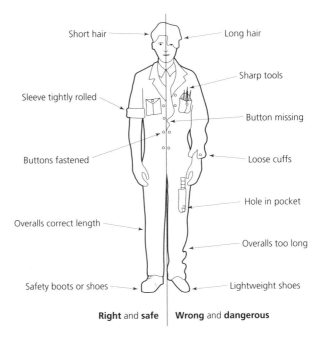

Short hair — Long hair

Sharp tools

Sleeve tightly rolled — Button missing

Buttons fastened — Loose cuffs

Hole in pocket

Overalls correct length — Overalls too long

Safety boots or shoes — Lightweight shoes

Right and **safe** | **Wrong** and **dangerous**

Long hair

- Long hair is liable to be caught in moving machinery such as drilling machines and lathes. This can result in the hair and scalp being torn away, which is extremely dangerous and painful. Permanent disfigurement will result and brain damage can also occur.
- Long hair is also a health hazard, as it is almost impossible to keep clean and free from infection in a workshop environment. Either adopt a short and more manageable head style or some sort of head covering that will keep your hair out of harm's way. Suitable head protection is discussed in Section 8.10.

Sharp tools

Sharp tools protruding from the breast pocket can cause severe wounds to the wrist. Such wounds can result in paralysis of the hand and fingers.

Buttons missing and loose cuffs

Since the overalls cannot be fastened properly, it becomes as dangerous as any other loose clothing and is liable to be caught in moving machinery. Loose cuffs are also liable to be caught up like any other loose clothing. They may also prevent you from snatching your hand away from a dangerous situation.

Hole in pocket

Tools placed in a torn pocket can fall through onto the feet of the wearer. Although this may not seem potentially dangerous, it could nevertheless cause an accident by distracting your attention at a crucial moment.

Overall too long

These can cause you to trip and fall, particularly when negotiating stairways.

Lightweight shoes

The possible injuries associated with lightweight and unsuitable shoes are:

- Puncture wounds caused by treading on sharp objects.
- Crushed toes caused by falling objects.
- Damage to your Achilles tendon due to insufficient protection around the heel and ankle.

Suitable footwear for workshop use is discussed in Section 8.13.

8.9.2 *Head protection*

As has already been stated, long hair is a serious hazard in a workshop. If it becomes entangled in a machine, as shown in Fig. 8.18, the operator can be scalped. If you wish to retain a long hair style in the interests of fashion, then your hair must be contained in a close-fitting cap. This also helps to keep your hair and scalp clean and healthy.

Fig. 8.18 *The hazard of long hair*

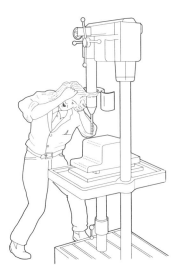

When working on site, or in a heavy engineering erection shop involving the use of overhead cranes, all persons should wear a safety helmet complying with BS 2826. Even small objects such as nuts and bolts can cause serious head injuries when dropped from a height. Figure 8.19(a) shows such a helmet.

Fig. 8.19 *Head and eye protection: (a) a typical fibre-glass safety helmet made to BS 2826; (b) plastic face safety visor for complete protection against chemical and salt-bath splashes; (c) transparent plastic goggles suitable for machining operations*

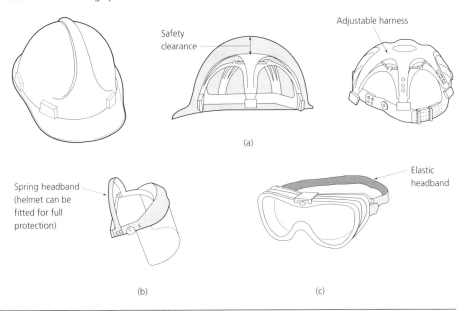

Safety helmets are made from high impact-resistant plastics or from fibre glass-reinforced polyester mouldings. Such helmets can be colour coded for personnel identification and are light and comfortable to wear. Despite their lightweight construction, they have a high resistance to impact and penetration. To eliminate the possibility of electric shock, safety helmets have no metal parts. The materials used in the manufacture of the outer shell must be non-flammable and their insulation resistance must be able to withstand 35,000 volts. A harness inside the helmet provides ventilation and a fixed safety clearance between the outer shell of the helmet and the wearer's skull. This clearance must be maintained at 32 millimetres. The entire harness is removable for regular cleaning and sterilising. It is fully adjustable for size, fit and angle to suit the individual wearer's head.

8.9.3 *Eye protection*

Whilst it is possible to walk about on an artificial leg, no one has ever seen out of a glass eye. Therefore eye protection is possibly the most important precaution you can take in a workshop. Eye protection is provided by wearing suitable goggles or visors, as shown in Fig. 8.19(b). When welding, special goggles (oxy-fuel gas welding) or visors (electric arc welding) have to be used. These have coloured lenses to filter out harmful rays. Eye protection when welding is discussed fully in *Fundamentals of Fabrication and Welding Engineering*.

Eye injuries fall into three main categories:

- Pain and inflammation due to abrasive grit and dust getting between the lid and the eye.
- Damage due to exposure to ultraviolet radiation (arc welding) and high-intensity visible light. (Particular care is required when using laser equipment.)
- Loss of sight due to the eyeball being pierced or the optic nerve cut by flying splinters of metal (swarf), or by the blast of a compressed air jet.

-

8.9.4 *Hand protection*

Your hands are in constant use and, because of this, they are constantly at risk handling dirty, oily, greasy, rough, sharp, hot and possibly corrosive and toxic materials. Gloves and 'palms' of a variety of styles and types of materials are available to protect your hands, whatever the nature of the work. Some examples are shown in Fig. 8.20. In general terms, plastic gloves are impervious to liquids and should be worn when handling oils, greases and chemicals. However, they are unsuitable and even dangerous for handling hot materials. Leather gloves should be used when handling sharp, rough and hot materials. **Never** handle hot workpieces and materials with plastic gloves. These could melt onto and into your flesh causing serious burns that would be difficult to treat.

Fig. 8.20 *Gloves suitable for heat treatment: (a) leather glove with reinforced palm – ideal for handling steel and sections; (b) gauntlet – available in rubber, neoprene or PVC for handling chemical, corrosive or oily materials; (c) heat-resistant leather glove – can be used for handling objects heated up to 360°C; (d) chome leather hand-pad or palm – very useful for handling sheet metal or glass; (e) industrial gauntlet – usually made of leather because of its resistance to heat; gauntlets not only protect the hands but also the wrists and forearms from splashes of molten salts and hot-quenching media.*

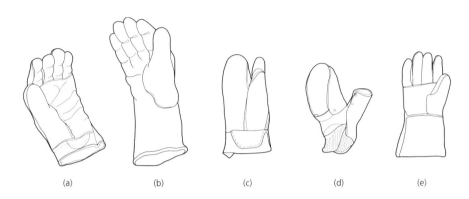

(a)   (b)   (c)   (d)   (e)

Where gloves are inappropriate, as when working with precision machines, but your hands still need to be protected from oil and dirt rather than from cuts and abrasions, then you should use a barrier cream. This is a mildly antiseptic cream which you can rub well into your hands before work. It fills the pores of your skin and prevents the entry of oils and dirt which could cause infection. The cream is water soluble and can be removed by washing your hands with ordinary soap and water at the end of the shift. Removal of the cream carries away the dirt and sources of infection.

Do not use solvents to clean your hands except under medical supervision. As well as removing oils, greases, paints and adhesives, solvents also remove the natural protective oils from your skin. This leaves the skin open to infection and can lead to cracking and sores. It can also result in sensitisation of the skin and the onset of industrial dermatitis.

8.9.5 *Foot protection*

The dangers associated with wearing unsuitable shoes in a workshop have already been discussed. The injuries which you can suffer when wearing lightweight, casual shoes are shown in Fig. 8.21. This figure also shows some examples of safety footwear as specified in BS 1870. Such safety footwear is available in a variety of styles and prices. It looks as smart as normal footwear and is equally as comfortable.

Fig. 8.21 *Safety footwear: (a) lightweight shoes offer no protection; (b) industrial safety shoe; (c) industrial safety boot*

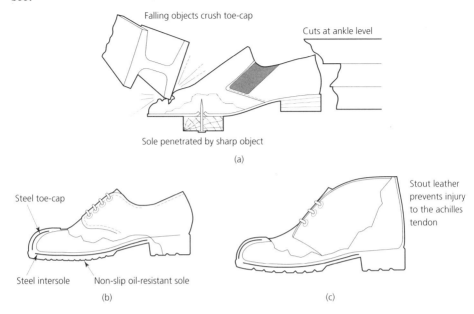

8.10 Hazards in the workplace

8.10.1 *Health hazards*

Noise

Excessive noise can be a dangerous pollutant of the working environment. The effects of noise can result in:

- Fatigue leading to careless accidents.
- Mistaken communications between workers, leading to accidents.
- Ear damage, leading to deafness.
- Permanent nervous disorders.

The level at which a noise becomes dangerous depends upon its frequency (pitch) and the length of time you are exposed to it. Noise is energy and it represents waste since it does not do useful work. Ideally it should be suppressed at source to avoid waste of energy and to improve the working environment. If this is not possible then you should be insulated from the noise by sound-absorbent screens and/or ear protectors (ear muffs).

Narcotic (anaesthetic) effects

Exposure to small concentrations of narcotic substances causes headaches, giddiness and drowsiness. Under such conditions you are obviously prone to accidents since your judgement and reactions are adversely affected. A worker who has become disorientated by the inhalation of narcotics is a hazard to himself or herself and a hazard to other workers.

Prolonged or frequent exposure to narcotic substances can, even in relatively small concentrations, lead to permanent brain damage and permanent damage to other organs of the body. Exposure to high concentrations can result in rapid loss of consciousness and death. Examples of narcotic substances are to be found amongst the many types of solvent used in industry. Solvents are used in paints, adhesives, polishes and degreasing agents. Careful storage and use is essential and should be carefully supervised by qualified persons. Fume extraction and adequate ventilation of the workplace must be provided when working with these substances. Suitable respirators should be available for use in emergencies.

Irritant effects

Many substances cause irritation to the skin both externally and internally. They may also sensitise your skin so that it becomes irritated by substances not normally considered toxic. External irritants can cause industrial dermatitis by coming into contact with your skin. The main irritants met with in a workshop are oils (particularly cutting oils and coolants), adhesive, degreasing solvents, and electroplating chemicals.

Internal irritants are the more dangerous as they may have long-term and deep-seated effects on the major organs of the body. They may cause inflammation, ulceration, internal bleeding, poisoning and the growth of cancerous tumours. Internal irritants are usually air-borne pollutants in the form of dusts (asbestos fibres), fumes and vapours. As well as being inhaled, they may also be carried into your body on food handled without washing or from storing noxious substances in discarded soft-drinks bottles without proper labelling. Many domestic tragedies happen this way.

Even the cutting oils used on machine tools can be dangerous if you allow your overalls to become impregnated with the spray. Figure 8.22 shows a factory worker fully protected against the cutting oil spray from a machine. These oils contain extreme pressure additives that can cause industrial dermatitis and even scrotal cancer if your overalls become heavily impregnated. Change your overalls regularly. The same gloves, goggles and apron would also protect you from hot and sharp metal chips and swarf. Such protective clothing should also be worn when handling corrosive and caustic chemicals and solvents.

Fig. 8.22 *Oil-proof protective clothing*

Systemic effects

Toxic substances, also known as systemics, affect the fundamental organs and bodily functions. They affect your brain, heart, lungs, kidneys, liver, your central nervous system and your bone marrow. Their effects cannot be reversed and thus lead to chronic ill-health and, ultimately, early death. These toxic substances may enter the body in various ways.

- Dust and vapour can be breathed in through your nose. Observe the safety codes when working with such substances and wear the respirator provided no matter how inconvenient or uncomfortable.
- Liquids and powders contaminating your hands can be transferred to the digestive system by handling food or cigarettes with dirty hands. Always wash before eating or smoking. Never smoke in a prohibited area. Not only may there be a fire risk, but some vapours change chemically and become highly toxic (poisonous) when inhaled through a cigarette.
- Liquids, powders, dusts and vapours may all enter the body through the skin:
  (a) directly through the pores
  (b) by destroying the outer tough layers of the skin and attacking the sensitive layers underneath
  (c) by entering through undressed wounds.

Regular washing, use of a barrier cream, use of suitable protective (plastic or rubber) gloves, and the immediate dressing of cuts (no matter how small) are essential to proper hand care.

8.10.2 *Personal hygiene*

As already stated above, personal hygiene is most important. There is nothing to be embarrassed about in rubbing a barrier cream into your hands before work, about washing thoroughly with soap and water after work, or about changing your overalls regularly so that they can be cleaned. Personal hygiene can go a long way towards preventing skin diseases, both irritant and infectious.

8.10.3 *Behaviour in workshops*

In an industrial environment horseplay infers reckless, foolish and boisterous behaviour such as pushing, shouting, throwing things and practical joking by a person or a group of persons. This cannot be tolerated. There is no place for such foolish behaviour in an industrial workplace. Such actions can distract a worker's attention and break his or her concentration. This can lead to scrapped work, serious accidents and even fatalities.

Horseplay observes no safety rules. It has no regard for safety equipment. It can defeat safe working procedures and undo the painstaking work of the safety officer by the sheer foolishness and thoughtlessness of the participants.

The types of accident caused by horseplay depend largely on the work of the factory concerned and the circumstances leading to the accident. Generally such accidents are caused when:

- A person's concentration is disturbed so that he or she incorrectly operates a machine or inadvertently comes into contact with moving machinery or cutters.

- Someone is pushed against moving machinery or factory transport.
- Someone is pushed against ladders and trestles upon which people are working at heights.
- Someone is pushed against and dislodges heavy, stacked components.
- Electricity, compressed air or dangerous chemicals are involved.

8.10.4 *Hazards associated with hand tools*

Newcomers to industry often overlook the fact that, as well as machine tools, badly maintained and incorrectly used hand tools can also represent a serious safety hazard. Unfortunately, the newcomer can be influenced by more experienced men and women – who should know better – misusing hand tools.

The time and effort taken to fetch the correct tool from the stores or to service a worn tool is considerably less than the time taken to recover from injury. Figures 8.23 and 8.24 shows some badly maintained and incorrectly used hand tools. Chipping screens, as shown in Fig. 8.25, should be used when removing metal with a cold chisel to prevent injury from the pieces of metal flying from the cutting edge of the chisel. For this reason, goggles should also be worn and you should never chip towards another worker.

Fig. 8.23 *Hand tools in a dangerous condition: (a) hammer faults; (b) chisel faults; (c) spanner faults; (d) file faults*

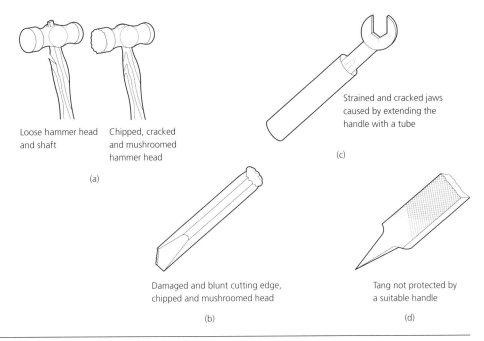

Loose hammer head and shaft

Chipped, cracked and mushroomed hammer head

(a)

Strained and cracked jaws caused by extending the handle with a tube

(c)

Damaged and blunt cutting edge, chipped and mushroomed head

(b)

Tang not protected by a suitable handle

(d)

Fig. 8.24 *Misuse of hand tools: (a) do not use an oversize spanner and packing – use the correct size of spanner for the nut or bolt head; (b) do not use a file as a lever*

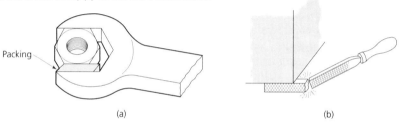

(a)   (b)

Fig. 8.25 *Use of chipping screen*

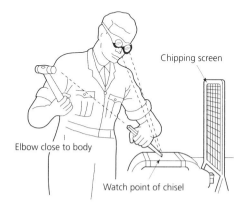

Chipping screen

Elbow close to body

Watch point of chisel

8.10.5 *Hazards associated with portable power tools*

In addition to maintaining portable power tools in good order, it is imperative that the equipment and, particularly the power lead, is electrically sound or a fatal electric shock can occur to the user. It is equally important that the compressed air hose supplying pneumatic tools is also in good order. When using electrically powered tools the following points should be observed:

- Be sure that the tool is correctly earthed or that it is 'double-insulated'. Earthing has already been considered in Section 8.4.
- Do not attempt to use faulty equipment. Faults in electrical equipment must be corrected by a qualified electrician. You must report the following unsafe conditions:
  (a) defective or broken insulation (tool and supply cable)
  (b) broken or defective plug
  (c) improperly or badly made connections to terminals
  (d) loose or broken control switch
  (e) incorrect voltage for the supply available
- Do not overload the motor, the heat generated will damage the insulation and cause excessive sparking at the brushes.

- Do not use portable power tools in the presence of flammable vapours and gases. Sparks from the equipment can cause a serious explosion and fire. Some electrical power tools are specially made so that they are safe to use when flammable substances are present. However, it is best to use pneumatically powered tools in such situations. Similarly, the process for which the power tool is required must not give rise to sparks. For example, a portable grinding machine must not be used in the presence of flammable substances.
- Although domestic power tools are designed to operate from the mains supply at the full mains voltage, industrial equipment is designed to work at a lower voltage to minimise the effect of electric shock. Such equipment is designed to operate at 110 volts and this lower voltage is supplied by an isolating transformer, as described in Section 8.4.

8.10.6 *Hazards associated with machine tools*

Metal-cutting machines are potentially dangerous. Tools designed to cut through solid metal will not be stopped by flesh and bone.

- Before operating any machinery be sure that you have been fully instructed in how to use it, the dangers associated with it, and that you have been given permission to use it.
- Do not operate a machine unless all the guards and safety devices are in position and are operating correctly.
- Make sure you understand any special rules and regulations applicable to the particular machine you are about to use, even if you have been trained on machines in general.
- Never clean or adjust a machine whilst it is in motion. Stop the machine and isolate it from the supply.
- Report any dangerous aspect of the machine you are using, or are about to use, immediately and do not use it until it has been made safe by a suitably qualified and authorised person.
- A machine may have to be stopped in an emergency. Learn how to make an emergency stop without having to pause and think about it and without having to search for the emergency stop switch.

Transmission guards

By law, no machine can be sold or hired out unless all gears, belts, shafts and couplings making up the power transmission system are guarded so that they cannot be touched whilst they are in motion. Figure 8.26 shows a typical transmission guard.

Sometimes guards have to be removed in order to replace, adjust or service the components they are covering. Before removing guards or covers, you must:

- Stop the machine.
- Isolate the machine from its energy supply.
- Lock the isolating switch so that it cannot be turned on again whilst you are working on the exposed equipment, and keep the key in your pocket.
- If it is not possible to lock off the supply, remove the supply fuses for the machine and keep these in your pocket until you have finished and the guards or covers are back in position.

Fig. 8.26 *Typical transmission guard for a belt-driven machine*

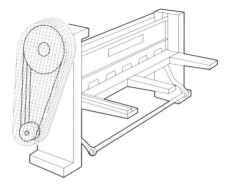

If an 'interlocked' guard is opened or removed, an electrical or mechanical trip will stop the machine from operating. This trip is only provided in case you forget to isolate the machine. It is no substitute for full isolation.

Cutter guards

The machine manufacturer does not normally provide cutter guards because of the wide range of work a machine may have to do.

- It is the responsibility of the owner or the hirer of the machine to supply his or her own cutter guards.
- It is the responsibility of the setter and/or the operator to make sure that the guards are fitted and working correctly before operating the machine, and to use the guards as instructed. It is an offence in law for the operator to remove or tamper with the guards provided.
- If ever you are doubtful about the adequacy of a guard or the safety of a process, consult your instructor or your safety officer without delay.

The simple drilling machine guard shown in Fig. 8.27(a) only covers the chuck and is only suitable for jobbing work when small-diameter drills are being used. The drill chuck shown in Fig. 8.27(b) is used for larger drills and for drills which are mounted directly into the drilling machine spindle. It covers the whole length of the drill and telescopes up as the drill penetrates into the workpiece.

Fig. 8.27 *Drill guards: (a) simple; (b) telescopic*

Figure 8.28(a) shows a horizontal milling machine guard. This is suitable for jobbing work where the setting of the guard has to be changed frequently and the machine is being used by a skilled operator. For production work the cutting zone is frequently totally enclosed, as shown in Fig 8.28(b) – particularly when the operator is not fully skilled.

Fig. 8.28 *Milling cutter guards: (a) for production milling; (b) for general machining when machine is used by a skilled operator (source: Silvaflame Ltd)*

(a)

(b)

Figure 8.29 shows a chuck guard for a centre lathe. This not only prevents the operator coming into contact with the revolving chuck, it also prevents coolant being thrown out by the chuck jaws. Since the guard will not close unless the chuck key has been removed, it is also a reminder to the operator to remove the key before starting the machine. Failure to remove the chuck key is a frequent cause of accidents and damage to the machine.

Figure 8.30 shows a fully guarded power press. Not only is the transmission mechanism and the flywheel fully guarded, but the work zone is also fully guarded. The guard fitted to the work zone has to be opened so that the tools can be loaded and unloaded. To achieve this safely an interlocked guard is used. This guard is interlocked with the clutch so that the

Fig. 8.29 *Lathe chuck guard (source: Silvaflame Ltd)*

machine cannot be started whilst the guard is open. It is also interlocked with a positive brake so that if a fault in the control mechanism causes the clutch to engage accidentally, the machine will stall and will not trap the operator.

Fig. 8.30 *Fully guarded machine – the press can only be operated with the 'gate' shut*

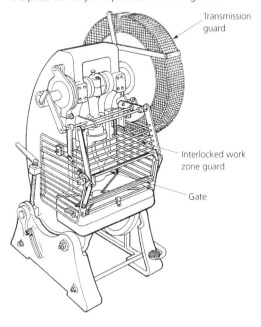

Transmission guard

Interlocked work zone guard

Gate

8.10.7 *Use of grinding wheels*

Figure 8.31(a) shows a double-ended, off-hand grinding machine of the type widely used in workshops for sharpening small cutting tools. Figure 8.31(b) shows how the tool rest, visor and guard should be set for safe operation.

Because of its apparent simplicity, the double-ended, off-hand grinding machine comes in for more than its fair share of abuse. A grinding wheel does not 'rub the metal off', it is a precision multi-tooth cutting tool in which each grain has a definite cutting geometry. Therefore, a grinding wheel must be selected, mounted, dressed and used correctly if it is to cut efficiently.

Fig. 8.31 *Safe use of the off-hand grinding machine: (a) the double-ended off-guard machine; (b) grinding wheel guard and work rest adjustment*

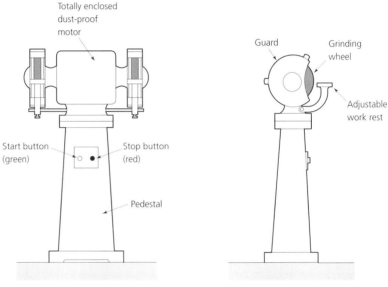

(a)

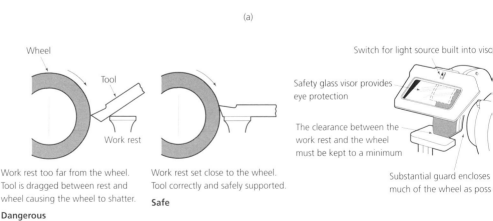

(b)

Further, a grinding wheel which is damaged, incorrectly mounted or is unsuitable for the machine, may burst at speed and cause serious damage and injury. Thus, the guard for a grinding wheel not only stops the operator coming into contact with the wheel, but it must also provide *burst containment* in case the wheel shatters or bursts in use – that is, the broken pieces of the wheel must be contained within the guard and not thrown out of the machine at high speed. For this reason grinding wheel guards are made very much stronger than most cutter guards.

The mounting of grinding wheels, except by a *trained and registered person*, is prohibited under the Abrasive Wheel Regulations 1970. The mounting and use of abrasive (grinding) wheels are considered in more detail in *Fundamentals of Mechanical Engineering*.

8.11 Loads and safety

In the engineering industry, loads are defined as heavy and cumbersome objects such as machines, large castings and forgings, heavy bar, sheet and plate materials, etc., which have to be loaded onto vehicles, unloaded from vehicles and moved within the factory itself. The movement of heavy loads involves careful planning and the anticipation of potential hazards before they arise. When moving such loads it is important that you use the correct handling techniques and observe the appropriate safety precautions and codes of practice at all times.

8.12 Manual lifting

8.12.1 *Individual lifting*

In the engineering industry it is often necessary to lift fairly heavy loads. As a general rule, loads lifted manually should not exceed 20 kg. Mechanical lifting equipment should be used for loads in excess of 20 kg. However, even lifting loads less than 20 kg can cause strain, and lifting loads incorrectly is one of the major causes of back trouble. The risk of personal injury and damage to equipment can be reduced by taking simple precautions before the lifting or handling operations begin. For example, if the load is obviously too heavy or bulky for one person to handle, you should ask for assistance. Even a light load can be dangerous if it obscures your vision as shown in Fig. 8.32. All moveable objects which form hazardous obstructions should be moved to a safe place before movement of the load commences.

As has already been stated, it is important to use the correct lifting technique. This is because the human spine is not an efficient lifting device. If it is subjected to heavy strain, or incorrect methods of lifting, the lumbar discs may be damaged causing considerable pain. This is often referred to as a 'slipped disc' and the damage (and pain) can be permanent.

Fig. 8.32 *Obstructions to safe movement must be removed*

Clear movable objects

The correct way to lift a load manually is shown in Fig. 8.33. You should start the lift in a balanced squatting position with your legs at hip width apart and one foot slightly ahead of the other. The load to be lifted should be held close to your body. Make sure that you have a safe and secure grip on the load. Before taking the weight of the load, your back should be straightened and as near to the vertical as possible. Keep your head up and your chin drawn in, this helps to keep your spine straight and rigid, as shown in Fig. 8.33(a). To raise the load, first straighten your legs. This ensures that the load is being raised by your powerful thigh muscles and bones, as shown in Fig. 8.33(b), and not by your back. To complete the lift, raise the upper part of your body to a vertical position, as shown in Fig. 8.33(c).

Fig. 8.33 *Correct manual lifting: (a) keep back straight and near vertical; (b) keep your spine straight; (c) straighten your legs to raise load*

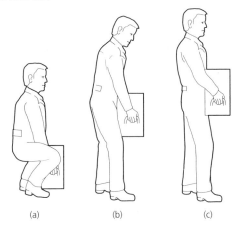

(a)   (b)   (c)

To carry the load, keep your body upright and hold the load close to your body, as shown in Fig. 8.34(a). Wherever possible hold the load so that the bone structure of your body supports the load, as shown in Fig. 8.34(b). If the load has jagged edges wear protective gloves, and if hazardous liquids are being handled wear the appropriate protective clothing, as shown in Fig. 8.34(c).

Fig. 8.34 *Correct carrying: (a) keep body upright and load close to body; (b) let your bone structure support load; (c) wear appropriate clothing*

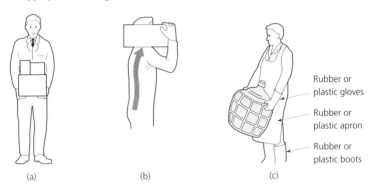

Rubber or plastic gloves

Rubber or plastic apron

Rubber or plastic boots

(a)   (b)   (c)

8.12.2 *Team lifting*

When a lifting party is formed in order to move a particularly large or heavy load, the team leader is solely responsible for the safe completion of the task. The team leader should not take part in the actual lifting but should ensure that:

- Everyone understands what the job involves and the method chosen for its completion.
- The area is clear of obstructions and that the floor is safe and will provide a good foothold.
- The members of the lifting party are of similar height and physique, and that they are wearing any necessary protective clothing. Each person should be positioned so that the weight is evenly distributed.
- He or she takes up a position which gives the best all-round view of the area and will permit the development of any hazardous situation to be seen so that the appropriate action can be taken in time to prevent an accident.
- Any equipment moved in order to carry out the operation is put back in its original position when the task has been completed.

This sequence of events is shown in Fig. 8.35.

Fig. 8.35 *Team lifting*

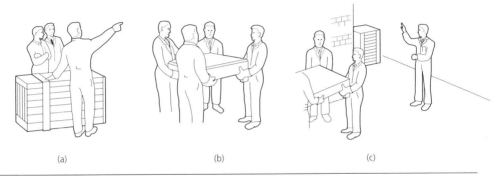

(a)   (b)   (c)

Loads which are too heavy to be lifted or carried can still be moved manually by using a crowbar and rollers, as shown in Fig. 8.36. The rollers should be made from thick-walled tubes so that there is no danger of trapping your fingers if the load should move whilst positioning the rollers. Turning a corner is achieved by placing the leading roller at an angle. As the load clears the rearmost roller, this roller is moved to the front, so that the load is always resting on two rollers, whilst the third roller is being positioned.

Fig. 8.36 *Use of rollers: (a) load is rolled forward on rollers 1 and 2 until it is on rollers 2 and 3; (b) when the load is safely supported on rollers 2 and 3, roller 1 is moved to the front ready for the next move*

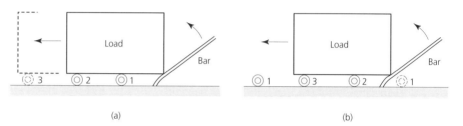

(a)   (b)

When lifting and moving loads manually, the following safety rules should be observed.

- Use the correct lifting technique – maintain your back straight and upright; bend your knees and let your legs and thighs do the work. They are much stronger than your back.
- Keep your arms straight and close to your body.
- Balance the load using both hands.
- Avoid sudden movements and any twisting of your spine.
- Take account of the centre of gravity of the load when lifting.
- Clear all obstructions from the area where the lifting and movement of loads is taking place.
- Take care to avoid injury to other people, particularly when moving long loads.
- Use gloves to protect your hands when moving loads with sharp or jagged edges.
- Use appropriate protective clothing when moving containers holding hazardous and toxic substances.
- Take particular care when moving materials which are wrapped and greased, they may slip through your grasp – wear safety shoes.
- Plan all stages of the move to avoid unnecessary lifting.
- Move all loads by the simplest method – simple is safe.

8.13 Mechanical lifting equipment

Mechanical lifting equipment can be classified according to the motive power used to operate it.

8.13.1 *Manual (muscle power)*

Examples of this type of equipment are shown in Fig. 8.37. Rope pulley blocks (snatch blocks) are light and easily mounted. However, the tail rope has to be tied off to prevent the load falling when the effort force is removed. Some rope blocks have an automatic brake which is released by giving the tail rope a sharp tug before lowering the load. They are suitable for loads up to 250 kg. Chain pulley blocks are portable and are used for heavier loads from 250 kg to 1 tonne. They also have the advantage that they do not run back (overhaul) when the effort raising the load is removed.

Fig. 8.37 *Manual lifting equipment: (a) rope pulley blocks (snatch blocks); (b) chain blocks (geared)*

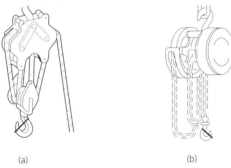

(a)   (b)

8.13.2 *Powered*

An example of a powered chain block is shown in Fig. 8.38. Such devices may be powered by electricity, compressed air or hydraulics. Internal combustion engined cranes are not

Fig. 8.38 *Powered lifting equipment*

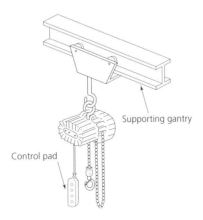

Supporting gantry

Control pad

usually used in factory environments because of the fumes. Powered lifting equipment is faster and can raise greater loads than manually operated chain blocks.

8.13.3 *Safety*

Only fully competent persons (i.e. trained and authorised) are permitted to operate mechanical lifting equipment. Trainees can only use such equipment under the close supervision of a qualified and authorised instructor. Even after you have been instructed in the use of lifting equipment, always make the following checks before attempting to raise a load.

- Check that the lifting equipment is suitable for the load being raised. All lifting equipment should be clearly marked with its safe working load (SWL).
- Never leave a load unattended whilst it is supported by lifting equipment and ensure that, before it is released from the lifting equipment, it is resting in a stable condition on a suitable support or supports.
- Hand chain: check the direction in which the chain must be pulled to raise or to lower the work as shown in Fig. 8.39(a).
- Cord control: when using cord-controlled power-operated hoists, check which cord raises the load and which cord lowers it, as shown in Fig. 8.39(b).
- Pendant switch control: if the pendant switch controls a travelling hoist, check the direction of travel as well as which press buttons raises and lowers the load, as shown in Fig. 8.39(c).

Fig. 8.39 *Check the operating procedure to raise, lower and move the load: (a) manual chain blocks – check which way the chain has to be pulled to raise or lower the load; (b) power hoist – check which cord raises and which lowers the load; (c) power hoist with power traverse – check all the pendant controls to raise, lower and move the load*

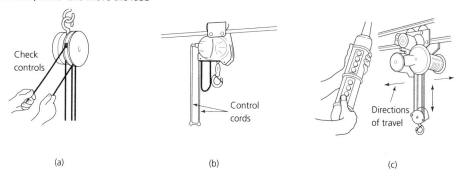

(a)   (b)   (c)

8.14 Use of lifting equipment

8.14.1 *Lifting a load*

Before lifting a load using a mechanical lifting device you should:

- Warn everyone near the load and anyone approaching the load to keep clear.
- Check that all slings and ropes are safely and securely attached to both the load and the hook.

- Take up the slack in the chain, sling or rope gently.
- Raise the load slowly and steadily so that it is just off the ground.
- Check that the load is stable and that the sling has not become accidentally caught on a part of the load incapable of sustaining the lifting force.
- Stand well back from the load and lift steadily.

8.14.2 *Traversing a load on a travelling crane*

Before traversing a load on an overhead travelling crane:

- Make sure the load does not have to pass over anyone working below.
- Check that there are no obstacles in the way of the crane and its load.
- Keep well clear of the load and move it steadily.
- Stop the crane immediately if anyone moves across the path of the load.

8.14.3 *Lowering a load*

Before lowering a load:

- Check that the ground is clear of obstacles and is capable of supporting the load.
- Place timbers under the load, as shown in Fig. 8.40(a), so that the sling will not be trapped and damaged. This will also facilitate the removal of the sling.

Fig. 8.40 *Care when lowering a load: (a) lower onto timbers; (b) guide by hand; (c) never work under a suspended load*

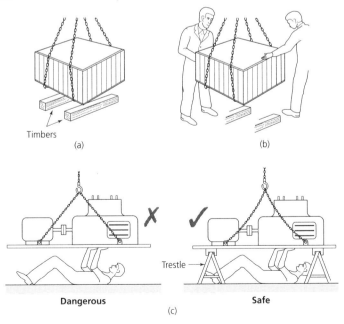

- Lower the load until it is close to the ground and then gently ease it onto the runners until the strain is gradually taken off the lifting equipment. It may be necessary to manually guide the load into place, as shown in Fig. 8.40(b), in which case safety shoes and protective gloves should be worn.
- Never work under a suspended load. Always lower the load onto suitable supports, as shown in Fig. 8.40(c).

8.14.4 *Hand signals*

If the hoist operator cannot see the load clearly, then the assistance of a signaller should be obtained. The signaller should be in such a position that he or she can see the load clearly and can also be clearly seen by the hoist operator. Both must be familiar with the standard code of hand signals, as shown in Fig. 8.41.

Fig. 8.41 *Standard code of hand signals for crane control*

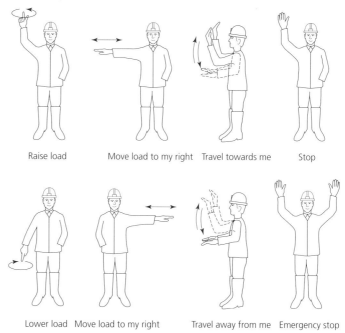

Raise load   Move load to my right   Travel towards me   Stop

Lower load   Move load to my right   Travel away from me   Emergency stop

8.15 Accessories for lifting gear

Hooks

These are made from forged steel and are carefully proportioned so that the load will not slip from them whilst being lifted. The hooks of lifting gear are frequently painted bright yellow to attract attention and to prevent people walking into them.

Slings

These are used to attach the load to the hook of the lifting equipment. There are four types in common use. They must all be marked with tags stating their safe working load (SWL).

- Chain slings (Fig. 8.42(a))   As well as general lifting, only this type of sling is suitable for lifting loads having sharp edges or for lifting hot materials.
- *Wire rope slings* (Fig. 8.42(b))   These are widely used for general lifting. They should not be used for loads with sharp edges or for hot loads; nor should they be allowed to become rusty. Further, they should not be bent round a diameter of less than 20 times the diameter of the wire rope itself.
- *Fibre rope slings* (Fig. 8.42(c))   Fibre rope slings may have eyes spliced into them or, more usually, be endless, as shown. They are used for general lifting, and are particularly useful where machined surfaces and paintwork have to be protected from damage.
- *Belt or strap slings*   Because of their breadth they do not tend to bite into the work and cause damage to the surface finish of the work. Rope and belt slings themselves must be protected from being cut or frayed by sharp edges, as shown in Fig. 8.42(d). This packing also prevents the fibres of the slings from being damaged by being bent too sharply around the corners of the object being lifted.

Fig. 8.42   *Types and care of slings: (a) chain sling; (b) wire rope sling; (c) fibre rope sling; (d) packing a sling*

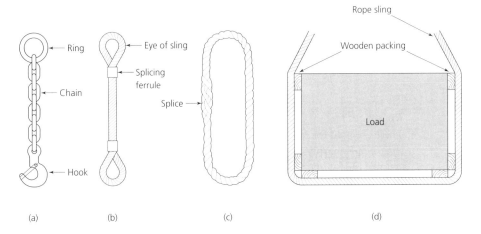

(a)   (b)   (c)   (d)

Care of slings

Wire rope and fibre rope slings must not be shortened by knotting since this twists and kinks the fibres causing them to fracture. Chain slings must not be shortened by bolting the links together.

Condition of slings

All slings must be checked before use for cuts, wear, abrasion, fraying and corrosion. Damaged slings must never be used and the fault must be reported.

Length of slings

Rope or chain slings must be long enough to carry the load safely and with each leg as nearly vertical as possible, as shown in Fig. 8.43(a). The load on a sling increases rapidly as the angle between the legs of the sling becomes greater. This is shown in Fig. 8.43(b).

Fig. 8.43 *Length of slings: (a) correct length; (b) effect of angle on the load*

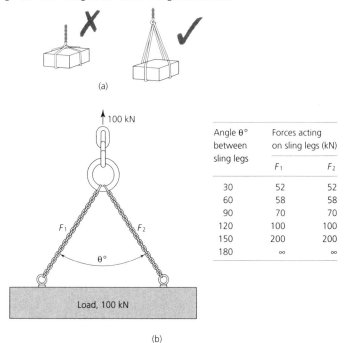

Angle $\theta°$ between sling legs	Forces acting on sling legs (kN)	
	F_1	F_2
30	52	52
60	58	58
90	70	70
120	100	100
150	200	200
180	∞	∞

Rings

These are used for ease of attachment of the sling to the crane hook. They also prevent the sling being sharply bent over the hook. Figure 8.44(a) shows a chain sling fitted with a suitable ring at one end. Figure 8.44(b) shows how a ring is used in conduction with a rope sling.

Eyebolts and shackles

Forged steel eyebolts to BS 4278 are frequently provided for lifting equipment and assemblies such as electric motors, gear boxes, and small machine tools. An example of the correct use of an eyebolt is shown in Fig. 8.45(a), whilst Fig. 8.45(b) shows how eyebolts must never be used. Forged steel shackles are used to connect lifting accessories together. In the example shown in Fig. 8.45(c), the eye of a wire rope sling is connected to an eyebolt using a shackle.

Special purpose equipment

Figure 8.46(a) shows how claws are used for lifting heavy barrels and drums. Particular care has to be taken when moving barrels or drums which are only partly filled. The movement of the barrel or drum can cause the liquid to surge about. This, in turn, changes the position

Fig. 8.44   *Use of rings: (a) with a two-leg chain sling; (b) with a chain sling*

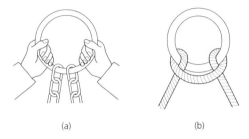

(a)   (b)

Fig. 8.45   *Use of eyebolts: (a) correct use; (b) incorrect use: (c) a shackle connects the eye of a sling to an eyebolt*

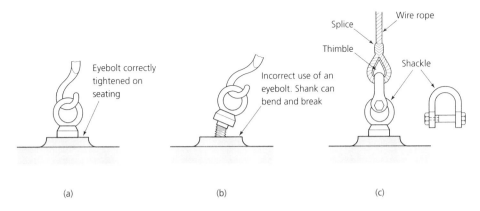

Eyebolt correctly
tightened on
seating

Incorrect use of an
eyebolt. Shank can
bend and break

Wire rope

Splice

Thimble

Shackle

(a)   (b)   (c)

Fig. 8.46   *Special purpose equipment: (a) use of barrel claws; (b) use of a spreader; (c) a plate hook*

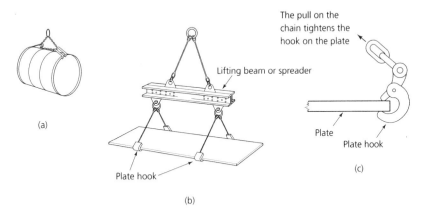

The pull on the
chain tightens the
hook on the plate

Lifting beam or spreader

(a)

Plate hook

Plate

Plate hook

(b)

(c)

of the centre of gravity and can cause the container to slip from the lifting equipment. If possible, completely fill or completely empty containers so that surging cannot take place. Figure 8.46(b) shows how plate hooks and a spreader are used. Spreaders are also used for lifting long loads such as bundles of tubing, girders, etc. Finally, Fig. 8.46(c) shows a plate hook in greater detail.

8.16 Useful knots for fibre ropes

It is useful to know about the knots that can be tied in fibre ropes can be useful when moving and securing loads. Knots must never be tied in wire ropes as they are not sufficiently flexible and permanent damage will be caused. Some widely used knots are as follows:

- *Reef knot*   This is used for joining ropes of equal thickness (Fig. 8.47(a)).
- *Clove hitch*   This is used for attaching a rope to a pole or bar (Fig. 8.47(b)).
- *Single loop*   This is used to prevent fibre ropes from slipping off crane hooks (Fig. 8.47(c)).
- *Half hitches*   Two half hitches can be used to secure a rope to a solid pole or for securing a rope to a sling (Fig. 8.47(d)).
- *Bowline*   This is used to form a loop which will not tighten under load (Fig. 8.47(e)).

Fig. 8.47   *Useful knots for fibre ropes: (a) reef knot; (b) clove hitch; (c) single loop; (d) two half hitches; (e) bowline*

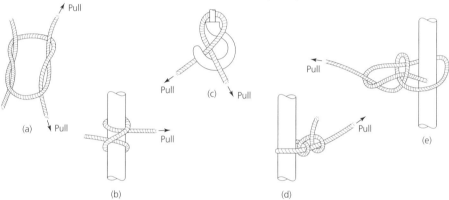

8.17 Transporting loads (trucks)

Various types of truck are used for transporting loads around workshops and factory sites. Only manually propelled trucks will be considered as power-driven trucks are beyond the scope of this book. The simplest sort of truck is the hand truck (sack truck) shown in Fig. 8.48(a). It uses the principle of levers to raise the load ready for wheeling away. Quite heavy loads can be moved quite easily with this type of truck.

Platform or flat trucks are used with various wheel arrangements so that they can be steered. The type shown in Fig. 8.48(b) requires the load to be placed over the wheels so that the truck is balanced for ease of movement. The type shown in Fig. 8.48(c) has a wheel at each end and the load does not have to be so carefully balanced. Only one end wheel is in contact with the ground at any one time. Also, the end wheels can slide, which facilitates steering. A heavier duty 'turntable' type truck is shown in Fig. 8.48(d). This has four wheels in the conventional position and the front wheels can be turned so that the truck can be steered.

Figure 8.48(e) shows a truck with a manually operated elevating table. The table is raised hydraulically by means of a double-acting hand pump. This type of truck removes the need for lifting the load, for example a heavy surface plate, from a bench down onto the truck, or from the truck up onto a bench. Finally, Fig. 8.48(f) shows a manually operated stacking truck for warehouse use. The load should always be lowered before the truck is moved. This lowers the centre of gravity of the truck and load and renders it stable and safe. Whatever type of truck you use there are two safety points you should observe.

- Stack the load on the trolley so that you can see where you are going in order to avoid a collision, particularly if people are working on ladders or steps in the vicinity.
- Balance the load so that it will not topple off the truck or overturn the truck when you are turning a corner.

Fig. 8.48 *Types of trucks: (a) sack truck; (b) two-wheel platform truck (balanced); (c) sliding wheel platform truck; (d) heavy duty, turntable-type platform truck; (e) elevating table truck; (f) stacking truck*

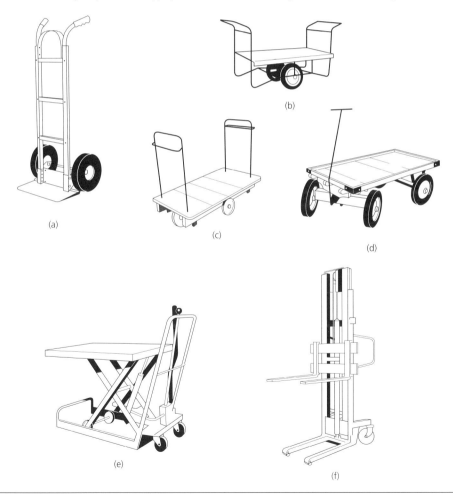

8.18 Inspection

It is a legal requirement under the Health and Safety at Work, etc., Act that all lifting equipment is regularly inspected by qualified engineers specialising in such work and that the results of such inspections are recorded in the register provided. If an inspector condemns any item of equipment it must be taken out of service immediately and either rectified or destroyed. If rectified, it must be reinspected and approved by a qualified inspector before being taken back into service. The inspector will, on each visit, also confirm the safe working load (SWL) markings for each piece of equipment. No new item of lifting equipment must be taken into service until it has been inspected and certificated.

8.19 Summary

8.19.1 *General rules for the observance of safe practices*

- *Keep alert*   A major source of accidents is fatigue. A tired worker can neither concentrate on the work in hand nor keep alert as to what is going on around him or her. Ensure you have regular and adequate meals and sleep.
- *Personal hygiene*   Wash regularly and use gloves or a barrier cream to avoid infections that can lead to industrial dermatitis and other skin diseases. Change your overalls regularly. Do not wear dirty and oily overalls. Keep your hair clean.
- *Protect yourself and other people*   Remember that careless working practices and horseplay can not only lead to your having an accident but it can also result in the injury or even the death of your workmates. Always obey the safety rules and practices as laid down in company policy.

Emergency procedures
- Make sure you know how to stop your machine quickly in the event of an emergency. Make sure you know where the emergency stop buttons for your shop are situated in case of an accident.
- Make sure you know the correct procedures to follow in the event of a fire: how to give the alarm; where the fire extinguishers are; how to select and use an extinguisher correctly; how to evacuate the premises in which you work; and how to call the professional brigade.
- Make sure you know the correct procedure for rendering first aid in the event of electric shock: how to send for assistance; how to turn off the supply or remove the victim from contact with the supply without endangering yourself; and how to carry out appropriate resuscitation techniques.
- Learn the basic first-aid procedures in the event of cuts, broken bones and loss of consciousness. Learn how to summon an ambulance or, in a large company, the company's medical staff.
- Report all hazards even if an accident has not occurred. Elimination of the hazard could prevent an accident or a fatality at some future date. After all, the victim might be you!

8.19.2 *Causes of accidents*

The main causes of accidents can be summarised as follows:

- Human carelessness leading to:
  - (a) improper behaviour and dress
  - (b) lack of training – never use equipment that you have not been trained on and with which you are unfamiliar
  - (c) lack of supervision and experience
  - (d) fatigue due to lack of sleep, inadequate rest periods, and irregular and inadequate meals
  - (e) drug taking, solvent abuse, consumption of alcohol prior to or during working hours
- Environmental causes of accidents such as:
  - (a) unguarded or faulty machines and tools
  - (b) inadequate ventilation, particularly if solvents are present as in paint spraying and when adhesives are being used
  - (c) poorly and incorrectly lit workplaces and stairs
  - (d) dirty and untidy workplaces with obstructed gangways and doorways
  - (e) overcrowded workplaces
  - (f) excessively noisy workplaces

8.19.3 *Accident prevention measures*

On the basis that prevention is better than cure, particularly where accidents are concerned, the following accident prevention measures should be observed:

- Always be on the look out for hazards.
- Eliminate any hazard as soon as it is discovered.
- If the hazard cannot be eliminated, then replace the process with one that is less dangerous, or place a guard round the hazard to prevent an accident.
- Provide personnel protection and ensure that it is used – for example, visors and goggles, protective headgear, gloves, ear protectors, safety shoes and boots, and respirators.
- Provide safety education and training with planned and frequent updating through retraining. Ensure that appropriate safety posters are displayed where they will be seen by all concerned and that they are replaced and updated regularly. Also ensure that they are drafted in all the languages used by the workforce.

8.19.4 *Attitudes to safety*

All employees, and especially young and new trainees (you!), need to develop personal attitudes to safety from the moment they enter the engineering industry. The management must provide the means through training and encouragement through company policy and example to ensure that all their employees are *safety conscious*.

All workers, young and old, new trainees as well as experienced workers, need to take a positive decision to act and work responsibly and safely in order to protect themselves, other people and the environment.

All workers in the engineering industry must have knowledge of:

- The dangers that can occur.
- The protection that is available.
- How to prevent accidents.

Examples of the above have been introduced throughout this chapter. The coverage is by no means exhaustive since every branch of the industry and every individual workplace and process has its own special hazards. However, by highlighting the more common hazards and how to avoid them you should realise the necessity to be always on the alert and you should have a better understanding of what to look out for and how to deal with new situations as they arise.

EXERCISES

8.1 *Health and Safety at Work, etc., Act and other important industrial legislation*
  (a) State:
      (i) what the initials HSE stand for
      (ii) what a prohibition notice means
      (iii) what an improvement notice means
      (iv) who issues the notices in (ii) and (iii) above
  (b) As an employee you also have duties under the Act. Copy out and complete Table 8.1 by writing brief comments regarding your duties in the circumstances listed in the table.

Table 8.1 *Exercise 8.1(b)*

Circumstances	Duties
You are uncertain how to operate a machine needed to complete your task	
You need to carry some sheet metal with very sharp edges	
You are working on site and you have mislaid your safety helmet	
You find that the belt guard has been removed from a machine you have been told to use	
You have spilt hydraulic oil on the floor whilst servicing a machine	
The earth wire has come disconnected from a portable power tool you are using	
Your supervisor has told you to clear up the rubbish left by another worker	
You find someone smoking in a prohibited area	

(c) Copy out and complete Table 8.2 by adding the name of the most appropriate industrial regulation(s) for each of the situations given.

Table 8.2 *Exercise 8.1(c)*	
Situation	*Appropriate industrial regulations*
Use of grinding machines and abrasive wheels	
Eye protection	
Electrical control equipment, use and maintenance	
Use of substances that can be harmful to health (solvents, etc.)	
Safe use of power presses	
Protection against high noise levels	
Safe use of milling machines	
Use of protective clothing	

(d) Copy out and complete Fig. 8.49 by stating:
   (i) the category of each sign (e.g. warning sign, mandatory sign, etc.)
   (ii) the meaning of each sign
   (iii) where each sign would be used

8.2 *Electrical hazards*
   (a) Explain why portable electrical equipment should be:
      (i) earthed unless it is 'double-insulated'
      (ii) operated from a low-voltage supply
      (iii) protected by an earth leakage isolator incorporating a residual current detector
   (b) When you are issued with portable electrical equipment from the stores you should make a number of visual checks before accepting and using the equipment. Describe these checks.
   (c) If the checks you made in (b) above showed the equipment to be faulty, what action should you take?
   (d) In the event of a workmate receiving a severe electric shock that renders him or her unconscious, what emergency action should you take?

8.3 *Fire hazards*
   (a) List **three** main causes of fire on industrial premises.
   (b) If you detect a fire in a storeroom at work, what action should you take?
   (c) Figure 8.50 shows various types of fire extinguisher:
      (i) state the types of fire upon which each extinguisher should be used and any precautions that should be taken
      (ii) state the colour coding that identifies each type of fire extinguisher.

Fig. 8.49 *Exercise 8.1(d)*

Sign	Meaning	Category	Where used

Fig. 8.50 *Exercise 8.3(c)*

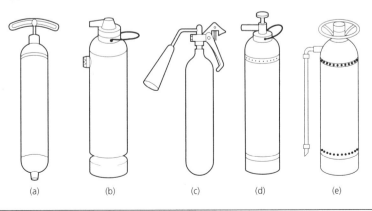

(a)   (b)   (c)   (d)   (e)

8.4 *Accidents*

    (a) Why should all cuts, bruises and burns be treated by a qualified first-aid person, and what risks does a well-meaning but unqualified person run in attempting first aid?

    (b) Apart from rendering first aid, what other action must be taken in the event of an accident?

    (c) State how you would identify a first-aid post.

    (d) State the first-aid treatment for the following injuries:

        (i) small cuts

        (ii) bruises

        (iii) minor burns

    (e) What action should you take if you come across someone who has received a serious accident (broken bones, severe bleeding, partial or complete loss of consciousness, etc.)?

    (d) Briefly describe the accident reporting procedures for your place of work or training workshop.

8.5 *Working environment*

    (a) Copy out and complete Table 8.3 by stating the type of working environment in which you would need to use the items of safety clothing and equipment listed in the table.

Table 8.3 *Exercise 8.5(a)*

Clothing/equipment	Situation/environment
Ear protectors	
Overalls	
PVC apron	
Leather apron	
Leather gloves	
PVC/rubber gloves	
Safety helmet	
Clear goggles	
Visor	
Barrier cream	
Safety boots	
Goggles with filter lenses	

    (b) With the aid of a sketch, explain what is meant by:

        (i) a transmission guard

        (ii) a cutter guard

(c) For each of the examples listed below, sketch an appropriate guard and state how it works:
   (i) travelling chip guard on a lathe
   (ii) milling cutter guard where the machine is being operated by a skilled operator
   (iii) a drill chuck guard
   (iv) a chipping screen
   (v) an interlocked transmission guard

8.6 *Lifting and carrying*
   (a) State the maximum recommended weight that may be lifted without the aid of mechanical lifting equipment.
   (b) With the aid of sketches, show the correct and incorrect ways to lift a load.
   (c) What precautions should be taken when carrying loads?
   (d) What precautions should be taken when moving a heavy load with a lifting team?
   (e) Lifting equipment should be marked with its SWL:
      (i) state what these initials stand for
      (ii) state how often lifting equipment needs to be tested and examined
      (iii) state the records that need to be kept

Index